ENERGY IN EUROPE

1945–1980

By the same author

THE COMMON MARKET

ENERGY IN EUROPE
1945-1980

W. G. JENSEN PH.D.

LONDON
G. T. FOULIS & CO LTD
1–5 PORTPOOL LANE E.C.1

First published 1967

S.B.N: 85429 074 5

Printed in Great Britain
at the University Printing House, Cambridge
(Brooke Crutchley, University Printer)

CONTENTS

To my Mother and Father

SECTION 1

THE POST-WAR FUEL SHORTAGE 1945–57 AND THE TWO-FUEL ECONOMY

The effect of the outbreak of the 1939–45 war on Europe's energy development

The outbreak of war in 1939 shattered the pre-war pattern of European coal exchange. The flow of coal from the major producing and exporting countries, that is, the United Kingdom, Germany and Poland, was disrupted or ceased altogether. Polish coals—and after the German military successes in 1940 in France, Belgium and Holland, coals from these countries also—were diverted from their normal end-use and transported to Germany to assist in building up the war machine. In the United Kingdom, the rapid successes of the German army, and the German conquests, brought the pre-war export trade to an abrupt halt and resulted in a temporary but short-lived surplus of coal. The immediate result was that many trained British miners were called up to join the army—a step that was to cause considerable difficulties within a few months.

In a survey completed in 1947 the Committee for European Economic Co-operation drew up a list of the principal effects of the war upon coal production in Europe. These were considered to be the adoption of ruthless exploitation methods with the aim of accomplishing maximum immediate results regardless of the future; neglect of normal maintenance work (even at the expense of safety) and of constructive development—a feature common to all producing countries during the war; loss in the efficiency of labour, partly due to organized resistance to German methods, and partly to lack of materials and the reduction of food and housing standards; and widespread use of inexperienced labour to bolster up production, for example, employment of prisoners-of-war and displaced persons on a large scale in the continental mines, by which expedient, wasteful in itself, production was in some cases expanded above pre-war levels. In the United Kingdom, as the war progressed, urgent steps were taken to rebuild the labour force which had been drastically reduced in 1940. Youths of 18, the famous Bevin Boys, who would otherwise have been drafted into the armed forces, were recruited and trained for work underground and were eligible for release at the end of the war, after their term of service had expired. During the latter war years special efforts were also made in the material field to speed up production, notably by the development of opencast mining for which purpose Lease-Lend equipment from the United States was brought over and set to work under the initial guidance of American technicians. On the Continent of Europe, towards the end of the war, when there was fighting on a large scale, widespread damage and destruction took place. In France the main coalfields suffered from bombardment and the deposits were flooded and the equipment pillaged. Many surface installations, for example, coke ovens in Holland, were also damaged or destroyed. In many cases reserve stocks at mines were used up or destroyed—in particular distributed stocks in the United Kingdom were heavily drawn upon to meet the requirements of the British and American Armies in France. In Germany also, as a result of the bombing campaigns and the advance of the Allied Armies into that country, the damage to mines was enormous (10% destroyed, 25% badly damaged).

The change over to war production had, inevitably, serious and far reaching effects upon the pattern of coal consumption. As a result severe restrictions on the use of solid fuel were imposed on domestic household consumption and all non-essential industries. In many of the occupied countries, wood was used extensively in place of coal—which was often virtually unobtainable—and during the winter of 1944 people in certain areas in the central and northern parts of Holland were reduced to burning the doors and woodwork of their own homes to provide some degree of warmth. Some idea of the degree of hardship involved in some of the other countries may be given by the fact that in France and Belgium domestic consumption per head per annum was cut during the period of the German occupation to approximately 15 to 20% of the pre-war level—and this sank still lower immediately after the liberation. A similar pattern developed in a number of non-occupied and non-coal-producing countries, for example, Switzerland, where consumption for domestic purposes and small industries had fallen by 1945 to 3% of the pre-war level.

The military campaigns of 1944 and 1945, which brought the war in Europe to a close, left coal production generally at a very low level and in some places brought it entirely to a standstill. Industry in most countries was similarly affected. Overland transport in continental countries was virtually non-existent—not only was there a shortage of fuel but in many cases lorries and vans had been commandeered and subsequently damaged or destroyed during the fighting. Stocks of fuel, both coal and oil, had been exhausted and the forced labour collected by the Germans and utilised by them in the pits and ancillary surface installations was rapidly dispersed. Practically no coal was available to meet domestic household requirements. In conditions such as these, the pre-war scale and pattern of trade in coal between the countries of Europe could not be quickly restored. Of the traditional European exporting countries, only the United Kingdom was able to supply substantial quantities of coal, and even these quantities were in the event drawn largely from stocks so that it was not possible to maintain exports at the initial level.

Faced with this situation, continental Europe made a tremendous effort to recover from the dislocation and chaos of war and rebuild or reconstruct its coal industry. Intensive efforts were made, often at the expense of other sections of the national economy. Good progress was made in re-manning the industry and in recovery of production. By May 1947, output of coal was twice as high as in July 1945 in Holland, almost double the level in Belgium and half as much again in France. In Germany the improvement was one of about 300 %. The figures (in indices) for these four Western countries, as well as the United Kingdom, are given in the table below:

Table 1. (1935–38: 100)

Country	Hard Coal Production		Workers on Colliery Books		Output per Manshift	
	July 1945	May 1947	July 1945	May 1947	July 1945	May 1947
Belgium	48	88	83	120	78	76
France	70	103	112	146	70	76
Netherlands	37	71	113	127	44	62
Western Germany						
Bi-Zone	18	51	52	89	51	56
Saar	28	80	36	87	73	81
United Kingdom	86	86	93	93	89	95

A report published by the U.S. Department of State[1] in 1948 gave as the principal measures that were taken to accomplish these remarkable results the priorities given to reconstruction of the mines and provision of equipment and special treatment to miners. In some countries (for example, Germany, Poland, Belgium) miners were given preferential ration scales as compared with the normal civilian ration. At the same time development plans, both short and long term, which were postponed during the war, were resumed and in all countries great efforts were made to redeem lost

[1] U.S. Department of State, Washington, Publication 2952: Committee of European Cooperation (European Series 29) published October 1947.

ground and to re-organise production along rational lines. These necessarily took different forms in different countries according to local conditions and resources. All these efforts, however, while wide in concept, were hampered to some extent by lack of financial aid and, above all, material resources. Special inducements were instituted for the encouragement of recruitment and for the promotion of output. These varied from one country to another, but in general they comprised wage increases, bonus schemes, schemes for food and consumer goods, special housing and transport schemes. In order to make good war-time losses of manpower (casualties, prisoners, etc.) as well as the exodus of forced labour, many countries had resorted to the use of prisoners of war, displaced persons and foreign free labour; labour of this character, however, took time to train and during the period of training did not, of course, produce at the full rate. Furthermore, the availability of prisoners of war could not be counted on for much more than a short period and for some countries this temporary expedient was merely a postponement of their labour problems.

Vigorous efforts were also made to rehabilitate the coal transport systems which, in the case of Continental countries in particular, had virtually been brought to a standstill at the time of the liberation. These efforts took the form notably of priorities for the transport of coal and the provision of wagons for coal transport at the expense of civil transport and other services.

In view of their own acute shortages of fuel it was only slowly—and often at very great hardship to themselves—that the pre-war major exporting countries were able to resume, albeit on a much lower scale, their export of coal to the importing countries. The United Kingdom, after having drawn heavily on its stocks to supply the allied Armies during the campaign of 1944 and 1945, resumed commercial exports of coal in 1946, although not in large quantities, averaging 330,000 tons a month between April 1946 and March 1947. This dwindled, however, during the latter part of the year as the fuel crisis in the United Kingdom worsened, and ceased entirely at the end of March. Exports from Western Germany were resumed in late 1945 in accordance with the directive addressed to the Commanders-in-Chief of the Allied Forces that exports from Germany to the liberated countries should be maximised. Owing to the slow recovery in output from the war-devastated mines, coal exports from Western Germany failed to meet expectations in 1946 and the early part of 1947 and only began to play a more important part in filling the energy gap towards the end of that year. Despite all these efforts, the extent of the general economic recovery in Europe could not have been maintained without substantial assistance from outside sources. In particular, without the substantial exports of American coal which flowed to Europe from mid-1945 onwards, industrial recovery in Western European countries would not have been achieved at the same pace. In 1945 United States exports to Europe totalled 519,000 tons. In 1946 they rose to more than 17 million tons—a rate of nearly 1½ million tons a month. Polish exports were also resumed towards the end of 1945 and in 1946 4·4 million tons of Polish coal were exported to Western European countries. In order to co-ordinate and allocate available supplies—taking into account the most pressing needs—discussions were begun, shortly before the end of the war, between a number of European countries and the United States with a view to setting up an international body whose main purpose would be to watch over the 'fair and equitable distribution' of available supplies of solid fuel. It was as a result of these discussions that the European Coal Organisation was set up and it was as a result of the recommendations from this organisation that American, German and other coals were allocated to a number of European countries.

The European Coal Organisation

The surrender of Germany and the end of World War II in Europe in May 1945 left Western Europe faced with an alarming energy problem. The large number of mines and surface installations which had either been completely destroyed or very severely damaged had reduced coal output—which before the war had constituted Europe's dominant source of energy, supplying over 90% of all her energy requirements—to a pitifully low level. Nor did the general circumstances of the time hold out much hope of improvement. In addition, however, to a low level of output in the liberated and war-devastated countries—this applied with equal force to the formerly occupied countries in Eastern Europe—the position was aggravated by a very heavy demand brought about by the virtual exhaustion of stocks in many of the former supplying countries (United Kingdom, Germany and Poland), the consumption requirements of the Allied Armies and the urgent needs of the reconstruction programme. In the spring of 1945 a joint Anglo-American Mission had examined the projects for the winter 1945/46 in the light of possible supplies of coal from Germany. On the basis of its findings, the Mission calculated that there was likely to be a gap between Europe's minimum requirements and the available supplies from all sources, including the United States, of about 3 million tons a month.

Faced with this situation it was evident that some central body would have to be set up with powers of direction and allocation in order to distribute the available supplies on a fair and rational basis. As long as military operations still continued, the allocation and distribution of coal were in the hands of the Supreme Headquarters Allied Expeditionary Force, but by the beginning of 1945, the end of the war and the consequent liquidation of SHAEF was in sight. Accordingly some members of the London Coal Committee (*i.e.* part of the war-time Anglo-American Combined Production and Resources Board) suggested to their respective Governments that some organisation should be established which would provide a meeting place for the coal exporting and importing countries and facilitate the task of working out satisfactory trading arrangements. The United Kingdom and the United States invited France and the U.S.S.R. to join in preliminary discussions. The U.S.S.R. later withdrew. The other three Governments eventually invited Belgium, Denmark, Greece, Luxembourg, the Netherlands, Norway, Turkey, the U.S.S.R., Czechoslovakia and Yugoslavia to join in such an organisation. With the exception of the last three countries all the others agreed. Arrangements were made for Czechoslovakia and Yugoslavia to attend the meetings of the organisation as observers. In this way, in May 1945, was set up on a provisional basis the European Coal Organisation. The Organisation was formally established some months later, on 1 January 1946, under an Agreement setting it up as an independent international Organisation to which the four founder Governments were signatories. (Poland and Czechoslovakia joined as full members during the course of 1946.) Its terms provided that it should remain in force initially for one year, but that it could be extended for a further period.

At the end of 1946 the member Governments unanimously declared that they wished the Agreement to be prolonged for a further year, and a Protocol for its formal extension was drawn up and signed by the member Governments. As it was already known at this time, however, that the Economic and Social Council of the United Nations was considering the establishment of an Economic Commission for Europe, the scope of which was expected to cover a large number of the functions and activities then falling within the compass of the E.C.O., a clause was inserted in the Protocol which provided that 'in the event of a new organisation being constituted on the initiative of the United Nations for the purpose of dealing with problems relating to fuel and power, the Council will consider, in conjunction with the new organisation, what steps should be taken for a transfer of the functions, assets and liabilities, personnel and archives of the European Coal Organisation, to the new organisation, and for a termination of the Agreement'. The Agreement defined the purpose of the European Coal Organisation to be to promote the supply and equitable distribution of coal and scarce items of mining supplies and equipment; to safeguard, as far as possible, the interests of coke producers and consumers; and to keep itself constantly acquainted with, and when necessary to discuss, the situation in regard to such supply and distribution, disseminate information in regard thereto, and make appropriate recommendation to the Governments concerned and to any other competent authorities.

When E.C.O. was first set up it had inevitably to take over the procedures used by SHAEF—allocations had been made on a monthly basis, with all transport of coal being under the direct control of SHAEF and liberated countries requiring mining equipment or supplies from the United States or the United Kingdom having to apply to SHAEF and being supplied through a system of requisitions. E.C.O. was therefore faced with the task of establishing new policies and methods of procedure to meet the changing circumstances in the period during which the Continent was practically demilitarised and the authority shifted from military to civilian responsibility.

In May 1946, the Organisation convened a major Coal Conference in Paris. The conference, which was attended by representatives from twenty-one countries and four international organisations, discussed at length the probable development of Europe's coal trade during the following year and found that unless exceptional measures were at once taken, the shortage of solid fuel in Europe over the ensuing twelve months was likely to be such as to cause widespread unemployment and seriously to retard post-war recovery. This catastrophe could be avoided only if effective steps were taken to ensure adequate food for miners, particularly underground workers, and to provide special incentives to attract recruitment and labour to the mines.[1] The Conference also called for special priority for the manufacture and distribution of mining equipment and mining supplies generally and for measures to ensure the availability of adequate facilities to lift and transport every ton of coal mined. These measures, it was stated, were not designed solely for the benefit of the coalmining industry, but were

[1] In France, an Order in Council of 14 June, 1946, granted a special status to everybody employed in the mines (*i.e.* better wages and holidays with pay); in the Netherlands, collaborators (with the Germans) were invited to work in the mines: 2,486 were so employed in 1947; in Poland, extra bonuses were paid and a special housing programme (prefabricated houses imported from Sweden and Finland) began in 1947; while in Belgium a number of measures were taken, including extra holidays with pay, exemption (or postponement) of military service, campaign by Fedechar in conjunction with the Belgian Government for miners to return to the pits, easy-term loans for miners' houses, and extra bonuses on certain terms to newly-recruited men who entered the industry as underground workers.

essential if dislocation of almost all industries, including food processing, were to be avoided; even so, they could not, in themselves, result in fulfilment of the minimum needs of Europe unless accompanied by rigid control of consumption of coal in Germany and maximum export of coal from that country.

The Conference went on to recommend that E.C.O. should make an urgent examination of the possibility of increasing coal production by the adoption of special methods of trading in coal, including bilateral agreements, as well as of the advisability of taking special account of certain end uses, such as steel production, when recommending the allocation of coal and coke.

By the second half of 1946 another problem was becoming increasingly evident and urgent, that is that the shortage of dollars would eventually effectively bring to a halt Europe's coal supplies unless she could once again become self-supporting. Supplies of American coal and their availability were improving but European countries were finding it increasingly difficult to make dollar goods available. While financial questions did not fall within the scope of the E.C.O., the situation towards the end of 1946 had become so serious that the Chairman addressed a letter to the Secretary General of the United Nations, drawing his attention to the gravity of the situation.

On 28 March 1947 the Economic and Social Council of the United Nations decided definitely to set up a regional Economic Commission for Europe. This Commission, at its first session in May, decided that the work of E.C.O., together with that of the Emergency Economic Committee for Europe and the European Central Inland Transport Organisation, should be absorbed—in the case of the E.C.O., not later than the end of 1947.[1]

E.C.O. assumed full responsibility for the allocation of export surpluses of coal available to Europe with effect from August, 1945. The July allocations had been prepared jointly by E.C.O. and SHAEF. At that stage the only claimant countries were Belgium, Denmark, France, Luxembourg, the Netherlands and Norway, and allocations were made upon an *ad hoc* basis after considering military and essential civilian requirements and any other relevant factors. This *ad hoc* basis of allocation gradually developed into a pattern representing percentage shares to each country of the total availabilities. These shares varied from month to month to meet any special circumstances, and the period of allocating on

[1] An important event in 1947 was the 'Moscow Agreement' by which the level of exports from Germany was fixed by a sliding scale in relation to production.

this system continued until November 1946. By then a new allocation system had been devised and came into operation. During the first period, from July 1945 to October 1946, almost 40 million tons of coal were allocated under the aegis of E.C.O. Under the new system of allocation, which sought to make allocations quarterly rather than monthly, the general principles governing allocations aimed at ensuring that the member countries received, firstly, their minimum basic relief needs or such percentage of them as availabilities allowed, and, secondly, supplementary allocations to provide an incentive for them to play their maximum part in the reconstruction of Europe in general and the promotion of coal production in particular.

The choice of the factors to be used in determining the allocations opened along these lines resulted inevitably in a number of difficulties that were only resolved by an elaborate system of compromises. Towards the end of the existence of E.C.O. a formula had, however, been more or less agreed, which took account of the following factors: pre-war consumption; war-damage allowance and hydro-power allowance; indigenous resources and bonus for production; deduction for low-grade fuel; exports. Each one of these factors was the source of considerable disagreement, and, to take only one example, the formula finally evolved to reach a compromise on the first of these factors was as follows:[1]

Monthly average 1935–38	A
„ „ 1929–38	13
Figure used in E.C.O. (for allocation purposes only	$\frac{A + 13}{2}$

[1] The formula had to be one which admitted the need for certain countries to receive some priority in the distribution of fuel, particularly where these countries were solely dependent upon imports. Therefore the first principle of the allocation formula ensured that: 'The countries poor in indigenous resources of coal shall receive first consideration whereby they may reach the same level of coal consumption as is enjoyed, on the average, by all countries out of their own production'. (*e.g.*, when in the fourth quarter of 1947 the indigenous resources of all producing countries participating in the allocation of availabilities amounted to 37 % of their combined pre-war consumption, no less than 24 % of the allocated coal was distributed among countries in order that all might attain such a percentage level of consumption.)

The second principle recognised by the formula was that: 'The balance of import availabilities thereafter remaining shall be divided in proportion to the amount by which countries will remain below the pre-war level of consumption.'

In general, these two principles ensured that all countries participating in the allocation machinery received a basic allocation of equal percentage satisfaction of their various national requirements, due allowance subsequently being made for the recognition of additional satisfaction to some countries for particular factors not existing in all countries, that is, production, war damage, hydro-electric power development.

This was because for some countries the most favourable period was 1935–38, while for others it was the longer period of 1929–38, and neither the one nor the other met with unanimous approval.[1]

During its lifetime, the Organisation recommended the allocation of a total quantity of 101·5 million tons of coal, while a further 14·1 million tons were distributed through the medium of bilateral agreements. This made a grand total, therefore, of 115·6 million tons:

(i) Coal allocated by E.C.O. (by country of origin)

	tons
Belgium	330,000
Germany	32,100,000
Netherlands	50,000
Poland	200,000
South Africa	900,000
U.K.	2,500,000
U.S.A.	65,400,000
Others	20,000
Total	101,500,000

(ii) Trade agreements taken into account involved the following supplying countries:

	tons
Austria	30,000
Belgium	2,367,000
Czechoslovakia	68,000
France	615,000
Hungary	7,000
Netherlands	678,000
Poland	9,543,000
Spain	16,000
Russian Zone of Germany	296,000
U.S.S.R.	344,000
Others	136,000
Total	14,100,000

The immediate post-war years

Coal

Despite all the efforts that were made in the fields of indigenous production, intra-European trade, outside assistance and co-ordination of distribution, it was not until 1948/49 that current availabilities began to catch up with consumption requirements. As a result, a number of countries were obliged to continue restricting household and non-essential industrial consumption and to concentrate the major part of their available supplies on essential industries. In general, first priority was granted to the requirements of public utilities (gas, electricity) and transport. Next in order of importance came the essential industries such as iron and steel, agriculture and (in certain cases) building materials. According to local circumstances and weather conditions, priority was also granted in some countries to domestic consumption over non-essential industries. It followed from this that, in periods of acute fuel shortages such as occurred in most of Europe

[1] It was difficult to select the basic year: 1939 was not acceptable, since most countries were then expecting war, with the result that their coal consumption figures were abnormal; 1937 and 1938 were objected to by others on the grounds of the effect of the then prevailing sanctions and political disturbances. The years 1930 to 1935 were also out because of the effect of the great depression and its aftermath.

during the bitterly cold winter of 1946/47, industrial activity was seriously curtailed and the progress of national recovery impeded. For example, in France and Denmark it proved necessary to order a complete stoppage of industrial activity for two to three weeks in the middle of the winters of 1945–46 and 1946–47, except in the case of continuous processes. In other countries the stoppage was confined to certain industries, textiles in Belgium, or in areas, *i.e.* South-Eastern England, where stocks at power stations ran out. Conditions such as these bore with particular weight on countries with little or no indigenous production and could only be alleviated in part by the system of international allocations designed to assure that the minimum requirements at least of the member countries of E.C.O. were met. The following table sets out the comparative consumption levels in 1946 and 1947 for all the member countries of E.C.O. as well as Western Germany in relation to the 1938 level.

Trends in levels of solid fuel consumption (including lignite) for E.C.O. area, 1946 and 1947[1] *(1938 : 100)*

Country	1946	1947	Country	1946	1947
Austria	67	77	Norway	65	74
Belgium	92	97	Portugal	76	89
Denmark	74	81	Sweden	41	62
France	84	93	Switzerland	38	58
Greece	40	34	Western Germany:		
Italy	55	93	British Zone	51	66
Luxembourg	71	83	French Zone	44	49
Netherlands	72	88	Saar	46	71
			United Kingdom	101	103

Another way of showing this comparison is in terms of *per capita* consumption. In 1938 the *per capita* consumption in the member states of E.C.O. and Western Germany was 2·2 tons per year as compared with 3·1 tons per year in the United States. In 1946 *per capita* consumption in the participating countries and Western Germany was only 1·6 tons per year, while that in the United States was nearly 4 tons per year.

It is, of course, quite impossible by statistical means alone to convey the full impact of such reduced availabilities in terms of human hardship. Only passing mention can be made here of the many expedients enforced by the general shortage of coal. Among examples which could be quoted were the large diversion of labour in Denmark from agricultural work to peat cutting and the burning of abnormal quantities of wood in Scandinavian countries during the winter of 1945/46 and 1946/47 owing to insufficient coal imports.

[1] Report by U.S. Department of State, *op. cit.* vol. II, p. 141.

Production Programme for the Four Years 1948–51

In 1947 the countries participating in E.C.O. and Western Germany agreed upon a number of production targets over the four-year period, ending in 1951.

Table 2.[1] *Coal production programme 1948–51, participating countries—including also Western Germany's hard coal production* (*in million metric tons*)

	1938	1948	1949	1950	1951
	483	414	446	482	511
As % of 1938:					
	100	86	92	100	106

The plan showed that the overall anticipated increase was a substantial one and that practically all countries shared in this increase. As compared with 1938 the production estimates for 1951 amounted to 112 % for the participating countries (105 %for Belgium, 131 % for France, 97 % for the Netherlands, and 108 % for the United Kingdom). For the Saar, the corresponding figure was 117 %, whereas for the Ruhr the figure was 87 %. If, however, the comparison was made with 1946, the appropriate percentages were as follows: participating countries, 131 %, Belgium 136 %, Netherlands 150 %, Ruhr 219 %, Saar 213 %, and the United Kingdom 129 %.

The United Kingdom forward production programme, covering both deepmine and opencast coal, provided for an increase in output from 1947 to 1951 from 199 million to 249 million (metric) tons. The National Coal Board's long-term programme provided for some twenty new sinkings and over twenty major construction schemes. In France, the Monnet Plan, which was adopted by the French Government at the end of 1946, set up a far-reaching programme for the French coal industry, which was regarded as a basic industry with full priority. This programme envisaged and was designed to bring about an increase in the total output from French mines from the 1947 level of a little over 50 million tons to 70 million tons in 1955. By 1951 output was expected to be up to 62·5 million tons. The most important element in the plan was the planned increase in output from the Lorraine coalfield. In Germany, production in the important Saar coalfield, which had already, in 1947, recovered to 80 % of the pre-war level, was planned to be materially expanded during this four-year period. Thus,

[1] Report by U.S. Department of State, *op. cit.* vol. II, p. 142.

the 1935–38 level of 12·5 million tons was expected to be exceeded by approximately one million tons, and by 1951 a level of output of 16·8 million tons was anticipated. In the Ruhr–Aachen coalfields there was a projected increase to 86 million tons in 1948, and to 121 million tons by 1951 (compared with 138 million tons in 1938). In addition it was planned to increase the level of brown coal production to 72 million tons in 1951—that is, 4 million tons more than the 1938 figure. The efforts required to bring about increases in output of this magnitude were expected to involve an increase in the effective labour force of some 180,000 men, and major repair programmes for housing and surface installations. New mechanised methods, including the development of the coal plough (which was an invention of the war years for mechanical coal-getting and loading, which was subsequently improved and developed) were also actively pursued. This programme, despite its tremendously ambitious character, was considered necessary and practical by the Bi-Zonal authorities who gave it priority. In the Netherlands, modern methods of mining (including new techniques) were extensively developed during the immediate pre-war years in order to meet intensive competition from the Ruhr. Thus, by 1939, a modern and efficient industry had been built up and practically all pits were working to full capacity. The target adopted for 1951 was about 13 million tons, or approximately the same level as in 1935–38, and it was considered doubtful whether, in view of the intensive exploitation of the South Limburg deposits, this figure could be much expanded. Even to attain it, it was considered to be necessary to pay higher wages and offer inducement to attract the men to the pits; and to give special priority for the manufacture and distribution of mining supplies. In Belgium, the Government laid down a production target by 1951 which was 1·5 million tons higher than the best pre-war figure in 1938. In the case of the smaller producers, the overall anticipated increase was about 50% over 1938. This reflected their anxiety to secure adequate supplies of fuel even at the cost of developing their own costly resources. Among these smaller producers, Greece, which had been driven to economise in the use of wood from its forests—many of which were destroyed or burnt during the war, made very great efforts to develop its production of lignite. The principal item of the reconstruction programme was concerned with the development, by opencast methods, of the Ptolemaid deposits in Central Macedonia, where there were 100 million tons of lignite. The lignite so produced was intended for making briquettes for the railways, and electric power generation. In Italy it was decided to increase substantially the Sardinian coal production, from 1·3 million tons in 1947 to 3 million tons by 1950.

In order to achieve this programme, it was intended to develop two additional mines and to build two housing programmes for the miners. An increase in output of lignite was also envisaged, although on a lesser scale than for Sardinian coal.[1]

Factors essential for the achievement of the production programme

One of the main problems facing the coalmining industries of Europe was that of obtaining the equipment required to refurbish the mines devastated by war. Total requirements of the participating countries for mine supplies and mining equipment over the four-year period in terms of sterling were estimated at over £510 million (covering all types of coalmining equipment, including underground machinery, locomotives and transport equipment, steam and electric generating plant, and machinery and equipment for use in preparation and cleaning plants). Of the total requirements listed by the major producing countries, about 75% represented requirements for normal maintenance, machinery replacement and the repair of war damage, while 25% only was required for development work, including large-scale reconstruction with the object of introducing new mining methods, the concentration and modernisation of existing pits and extensive new sinkings.

At the same time, all countries intensified the measures already taken to attract labour to the mines. The average age of the miners was comparatively high, and consequently a high rate of wastage was to be expected. Furthermore, prisoners of war and foreign labour previously employed compulsively in the mines were due for repatriation, and their loss would not be balanced by normal intake. It was therefore necessary that every possible step should be taken so that new labour could be recruited in a constant stream before there was a gap between the time that the total labour force was reduced and manpower economy could be effected by the introduction of mechanisation and up-to-date mining methods. In France the number of workers employed underground as at August 1947 (218,000 total), still included 36,000 prisoners of war. It was envisaged that the total labour force would be reduced to 210,000 in 1948, and 196,000 in 1951, in spite of an energetic recruiting drive in France itself, as well as French North Africa, Italy and Western Germany (notably among displaced persons). In the United Kingdom, an increase in the labour force from 719,000 in mid-summer 1947, to 740,000 by the end of 1951, was expected. It was obvious that unless urgent steps were taken to ensure that there would be adequate labour for the mines, the actual production pro-

[1] *Ibid.* pp. 143–46.

grammes in the various countries would be prejudiced. It was, however, not possible to find additional skilled mine-workers in the participating countries and it was therefore necessary for countries needing such manpower to recruit and train unskilled labour drawn from among those attaining the necessary physical standard.[1]

Special problems of Western Germany

In order to understand what took place in Western Germany after the collapse, and how vast a problem confronted the Allies in their effort to restore coal output, we must try to give a picture of the general situation.

The main coal-producing areas of Western Germany are, in order of importance, the Ruhr hard coalfield, the Saar hard coalfield, and the Cologne brown coal deposits. The latter were used mainly by local power-stations and for brown coal briquette production. The Ruhr hard coalfield produced, before the war, about 126 million tons per year. It constituted one of the largest deposits of coking coal in Europe, around which an industrial complex of the highest importance had been built. By 1945 the greater part of the vast industrial area had been reduced to rubble; transport was at a standstill and the difficulties to be surmounted before coal production could be effectively resumed were stupendous. [Thus, for instance, out of the 172,000 dwelling units in the Ruhr in 1939, 150,000 were destroyed or damaged.] One year later output had risen to a rate of nearly 60 million tons, and by July 1947, to over 75 million tons. Bearing in mind the appalling conditions at the outset of the occupation, this represented a very considerable achievement.

With regard to manpower it was estimated that—taking account of the high labour turnover—if the number of effective underground manshifts was to be raised by 40,000 to 50,000 per day during the 1948–51 period, an additional 180,000 men would be required and intensive recruiting drives were set in motion. Inducements for higher output were offered in the form of 'points' or coupon schemes related to individual attendance. (The goods provided included bacon, sugar and schnapps.)

Total solid fuel requirements

These estimated requirements show an increase from 471 million tons in 1947 to 609 million tons in 1951. This was due to the general increase in the level of economic and industrial activity in all countries, which sprang in part from the effort all were making to achieve rehabilitation and reconstruction over the period, and in part also from the more normal

[1] *Ibid.* pp. 149–50.

and permanent trends in European economic development (including the results of pursuing the policy of full employment, adopted by most countries). A typical example of the latter trend was the steady increase in the consumption of electricity in all countries. At the same time, domestic consumption everywhere had been drastically reduced and was not expected to have reached the pre-war level even in 1951.

Table 3.—*Consumption Requirements of Solid Fuel (including Bunkers) for E.C.O. area, 1947–51 (in million metric tons)*

1929	1938	1947	1948	1949	1950	1951
569	552	471	530	555	585	609

Import requirements

In the plans drawn up by E.C.O. it was envisaged that West Germany and the United Kingdom would provide substantial quantities of coal. Thus, in the case of the United Kingdom, exports to the participating countries were expected to reach 22 and 29 million tons respectively in 1950 and 1951. (These figures were never approached in the event, owing to failure to increase production at the rate forecast and the rapid rise in home demand.)

Table 4. *Solid Fuel Import Requirements of E.C.O. area, 1948–51 (in million metric tons)*

	Production less exports	Consumption	Import Requirements
1948	475	533	58
1949	508	558	50
1950	546	588	42
1951	575	612	37

Since the possibilities of providing supplies from the participating countries and Western Germany had been fully considered, there was no other recourse than to call on outside suppliers. The United States was considered to represent an abnormal source of supply, because of the huge distance involved and the consequently high freight rates. Poland, which after the war controlled both the Upper and Lower Silesian coalfields—important pre-war sources of supply to Western Europe—originally expected to increase its exports of coal to the participating countries from 8 million tons in 1947 to 30 million tons by 1951—but these figures, as in the case of the United Kingdom, were not, in fact, achieved. Imports from all other non-participating countries (*i.e.* South Africa, Hungary) were not expected to exceed about 1 million tons per year.

Table 5. *Assessment of Extra-European Solid Fuel Import Requirements of E.C.O. area, 1948–51 (in million metric tons)*

	Import Requirements	Imports from Poland and other countries	Import Requirements from United States
1938	10	10	—
1948	58	17	41
1949	50	25	25
1950	42	28	14
1951	37	31	6

The salient point which emerges from this table is the extent of the dependence of the European economy on extra-European sources of solid fuel supplies during the first half of the four-year period 1948–51 despite the tremendous efforts all the major coal-producing countries, including West Germany, intended to make in order to maximise output in the shortest possible time. By 1951, however, it was expected that it would be possible to dispense almost entirely with extra-European aid in terms of solid fuel imports. Thus, after 1948, the requirements from the United States were expected to fall in direct ratio to the planned growth of European coal production, and the expanding scale of United Kingdom, German and Polish exports to other European countries. It is, perhaps, worth noting, that even if it seemed possible to provide for all requirements as early as in 1948, albeit with some difficulty in that year, there remained a problem of quality: there was a shortage of good quality hard coke for the steel industry, i.e. in the figure of 41 million tons required from the United States were included 7 million tons of good quality coking coal, and some experts expressed very considerable doubts already in 1947 whether Europe would ever be able to meet its requirements of coking coal from its own resources. It is doubtful, however, whether even they foresaw the tremendous upsurge in European steel production which was to make imports of coking coal from the United States a permanent feature of the Western European energy scene.

Oil

With the object of reducing the heavy import demand for oil, and for other general economic reasons, substitute fuels from indigenous production, such as alcohol, benzole and tar oil, were already before the war being blended with petroleum products in some countries, but the quantities—as in the case of indigenous crude oil—were comparatively small. The synthetic production of liquid fuels from coal was of greater importance and Germany did, in fact, build up a large 'oil from coal' industry in the years before the war. The quantity of indigenous fuels—from natural crude and shale oil, synthetic production, and substitute fuels—used in the participating countries and West Germany during 1938 was estimated at about 2·2 million tons, representing 7% of the total requirements in that year. Imports of petroleum into the participating countries and West Germany during 1938, that is, excluding indigenous supplies, amounted to approximately 32·6 million tons, comprising 13 million tons of crude oil for processing in European refineries, and 19·6 million tons of finished products.

Prior to 1939, the method of supplying European countries with their oil requirements was primarily determined by commercial factors. The tendency in the oil industry was more to conduct refining in large units near the source of the crude oil rather than to transport the crude oil to the markets and refine it there. This was due mainly to the advantages of being able to use natural gases as refinery fuel, the greater ease in transporting a finished product, and the obvious economic advantages of operating one big refinery rather than a number of smaller ones. These and other allied considerations tended to confine the refinery industry to the areas where the crude oil was produced until, in the immediate post-war years, the exploitation of the enormous Middle East oilfields and the simultaneous expansion of demand for oil products in the United States itself resulted in the establishment of a large and growing refinery industry, first in the Middle East, and later in Western Europe. The phenomenal post-war expansion of the oil industry, which in the United States began as early as 1946, did not really get under way in Europe until after the Korean War.

The impact of war caused no fundamental alteration in the structure of the oil industry in the participating countries, the only important change being in Austria, where production of crude oil was increased appreciably by the Germans. On the other hand, refining plants and distribution facilities—particularly in France, Germany, Italy and the Netherlands—suffered very heavy damage. A large number of main storage tanks were either completely destroyed or severely damaged: many rail tank-cars and road lorries engaged in the petroleum trade were lost and refinery capacity was substantially reduced. During the war civil consumption of oil in occupied Europe and in the neutral countries was gradually reduced to a very low level as a result of which recourse was made to uneconomical (and unsatisfactory) substitute fuels such as producer gas, wood, peat, etc. The sources of supply to unoccupied Western Europe underwent drastic changes as the war proceeded. With the closing of the Mediterranean, the haul from the Persian Gulf became so long that it had to be eliminated in order to make the best use of the available tanker tonnage. In the end imports were shipped almost exclusively from the U.S. and the Caribbean area. This arrangement, in addition to increasing the carrying capacity of the tanker fleets, also had the advantage of enabling large quantities of oil to be supplied from the Persian Gulf to the U.S. Navy in the Pacific.

Table 6. *Crude Oil Consumption* (*in million metric tons*)

Country	1938	1946	% of 1938
France	6·82	2·79	40·9
Italy	1·50	0·05	3·3
Netherlands	0·70	0·45	64·3
Western Germany	1·54	0·50	32·4
Total	10·56	3·79	35·9

The search for oil in the participating countries—held up during the war in occupied countries with the exception of Austria—was renewed following the end of hostilities. In addition to the increased production in Austria, there were already then satisfactory prospects for the Netherlands; but results elsewhere in France, Italy, and the United Kingdom remained disappointing. The repair of war damaged refineries and storage tanks and the reinstitution of adequate transport and handling facilities was pushed on with all possible speed. Actual accomplishment would probably have been far greater had it not been for the acute shortage of materials such as steel. Despite all the adverse factors, however, a good recovery was made by some countries from the chaotic conditions that prevailed at the time of cessation of hostilities. France and the Netherlands by intensive efforts succeeded in repairing and operating by 1946 40·9% and 64·3% respectively of their pre-war refining capacity. On the other hand, in West Germany and Italy—the two other areas where an important local refining industry suffered severe material damage—recovery was slower, particularly in the case of Italy where in 1946 the crude oil throughput amounted to only 3·3% of the pre-war level.

It is worth recalling that, in those participating countries where UNRRA operated, petroleum products formed an important part of the programme of relief and rehabilitation. The position in 1947–48 broadly corresponded in the participating countries and West Germany with pre-war conditions. There was, however, relatively less refining of crude oil; consequently, a greater proportion of the requirements had to be imported in the form of finished products—at a greater cost, of course, in foreign currency. This was a direct result of war damage to European refineries.

The fact that the 1947 dollar bill for oil supplies and equipment was appreciably higher than it was before the war was due not to any significant change in the pattern of European oil economy, but rather to a greatly increased demand for oil products which was related to the shortage of solid fuels; a marked increase in the prices of oil products and oil equipment; the need for buying a higher proportion of finished products due to reduced refining capacity in the participating countries and West Germany; the loss of Rumanian oil supplies; and, finally, the purchase of abnormal quantities of equipment and material for the rehabilitation of damaged plant and for the building of new installations.

Oil Requirements, 1948–51

It was estimated that by 1951 total requirements would be about 59% higher than in 1947. This large-scale expected increase was due to the large-scale conversion of industrial plants from coal to fuel oil and the installation of diesel-driven generators due to the shortage of coal and electrical power. The coal which was thereby released was to become available for other purposes and these conversions were therefore regarded as important factors in assisting the general recovery of the European economy. (Under the then prevailing conditions, it was less costly to import fuel oil from whatever source than to bring in the equivalent quantity of American coal when account was taken of the greater calorific value of fuel oil.) Other factors which resulted in this rapid increase were the fact that actual fuel requirements for inland trade consumption, which in 1938 amounted to only 4·8 million tons, had risen to 12 million tons by 1947 (and were expected to rise to 28 million tons by 1951); the rate at which agriculture was being mechanised, resulting in a demand for additional petroleum products, particularly power kerosene; an increase in gasoline and diesel oil consumption for road haulage purposes—this increased requirement was, to a considerable extent, a direct consequence of damage to and inadequate maintenance of, alternative means of internal transport during the war—and the effect of population increases.

Table 7.[1] *Oil Requirements in E.C.O. area, 1938–51 (in million metric tons)*

Year	Refinery Output	%	Imports of Finished Product	%	Total Requirements	%
1938	13·7	40	19·6	60	33·3	100
1946	8·0	24	25·2	76	33·2	100
1947	12·0	28	31·1	72	43·1	100
1948	15·7	30	37·3	70	53·0	100
1949	18·8	32	40·3	68	59·1	100
1950	23·6	37	41·0	63	64·6	100
1951	30·7	45	38·0	55	68·7	100

Electricity

Before the war, most of the participating countries had sufficient generating plant to supply almost the whole of their own electricity requirements. Each of these countries also had sufficient spare plant to ensure an adequate and regular supply of electricity at normal frequency and voltage. Interchanges of electric power between different countries were therefore relatively small, except in the cases of Switzerland and Austria. The abundant water power resources in these countries made the output of electric power an attractive proposition and they supplied approximately 20 and 10% respectively of their total production to other countries.

During the years immediately before the war, the power requirements of the participating countries and West Germany increased continuously. In 1938 net production was 130 milliard kilowatt/hours and the total output capacity was 39 million kilowatts, of which 14 million kilowatts was hydro-electric plant. This production was equivalent to about 530 kilowatt hours per head of population per annum, which compares with a corresponding figure for the same date in the United States of nearly 900 kilowatt/hours.[2] Inevitably, one of the first results of the war was the slowing down or the complete abandonment of long-term programmes for new generating plant in favour of more urgent war needs. In addition, due to a variety of causes, the output capacity available from the existing plant was considerably reduced, in some cases permanently and in other cases temporarily until repairs could be made. The principal causes were: damage to plant by bombing and other military action; damage to plant due to the 'scorched earth' policy; danger to plant by sabotage; removal of plant by the occupying power; insufficient maintenance due to shortage

[1] *Ibid.* p. 187. [2] *Ibid.* p. 168.

of staff and materials and the necessity of abnormally long periods of operation; damage to plant by use of unsuitable fuel; damage to plant due to overload, operation at abnormally low frequencies and voltages, severe earth faults due to frequent war damage to transmission equipment, etc. (exhaustion of coal stocks at generating stations). The net result of these conditions was that the rate of installation of new plant and repair of existing plant only slightly exceeded the rate of reduction in output capacity of existing plant. Thus, the total generating plant capacity in the participating countries and West Germany increased by only 8% between 1938 and 1946. But as the Report by the U.S. Department of State revealed, 'while there was a halt in the installation of new generating plant, the installation of new consuming plant (motors, furnaces, etc.) continued in essential war industries and to a lesser extent in offices and homes. It was estimated that the increase in requirements in the eight-year period 1938–46 would have been about 50%, but of course these requirements could not be met, since no corresponding increase had taken place in generating plant capacity. When all spare plant had been brought into service, it thus became necessary to restrict requirements to the supplies available, by rationing, disconnection of supplies and other methods.'

'When the war ended, the most pressing need was to repair the damaged power stations and transmission systems in order to meet the essential needs of the population. This was done, first by makeshift, then by more permanent repairs. Efforts were also made to obtain the largest possible power supplies from Germany by the construction of suitable interconnectors. At the same time, the participating countries helped each other as far as possible, both by the supply of electric plant and by each country making available to its neighbours the maximum possible exports of electric energy. The United States also helped in many ways, in particular by putting floating power stations at the disposal of Belgium and the Netherlands. Moreover, most of the participating countries were represented on the Public Utilities Panel of the Emergency Economic Committee for Europe (subsequently replaced by the Power Committee of E.C.E.) and there established contacts and made arrangements for mutual assistance where this could be done. This common effort was, however, often seriously hampered by the chronic shortages of fuel, raw materials and labour. The production of the power stations could not keep pace with the demand. As a result, the frequency and voltage of the supply system often fell, sometimes by more than 10%, with consequent detrimental effects. To meet these conditions, a number of emergency measures were taken. Economy campaigns were undertaken to limit consumption to the minimum essential quantity (industrial and commercial users took over ¾ of the total supply in the participating countries and West Germany, and direct approach was made to them by the appointment in many countries of fuel economy "wardens" or "officers" in factories and workshops; domestic consumers, who took less than one-quarter of the total supply, were approached by broadcasting and other national publicity media); and arrangements were made to adjust industrial working hours in order to spread the electrical load more evenly throughout the twenty-four hours of the day and thus reduce the peak demands: this enabled the maximum possible output to be obtained from the limited plant capacity available; the consumption of electrical energy (kilowatt/hours) had to be rationed in many countries although this inevitably caused serious interference with industrial production and considerable hardship for the general public; plant maintenance was reduced and often postponed and the generating plant was kept in operation as long as possible; and, finally, programmes for the rapid construction of diesel-driven generators were adopted (especially in the United Kingdom) and coal-fuel boilers were converted so as to burn oil (especially in France and the Netherlands) in order to reduce the demands on the limited supply of solid fuel and in some cases to obtain additional output from existing boiler plant. Even these stringent emergency measures sometimes proved to be insufficient, and in addition, particularly during the winter months of 1945/46 and 1946/47 and 1947/48 (in some cases), it became necessary to disconnect completely the supply to certain consumers during the peak load hours. Such disconnections were usually made according to a predetermined rota so as to spread the hardship as evenly as possible over all sections of the community.'[1]

Programme for years 1948–51

The increase in productive capacity in the participating countries and West Germany during the last full pre-war year, 1938, was only about 1·5 million kilowatts. The programme of construction put in hand for the period 1948–51 therefore represented a very good effort to obtain a balance between productive capacity and requirements at the earliest possible date. In this connection, where existing turbine capacity was greater than the associated boiler capacity, arrangements were made, wherever possible, to obtain additional boilers, etc. The relatively small figure (1·9 million kilowatts) for 1947 only serves to emphasize the long period needed for the construction of new power-stations. In the following

[1] *Ibid.* pp. 170–71.

years the installation of new plants increased at a much more rapid rate (*i.e.* three to four times the 1938 figure). Many countries decided to assign very high priority to the manufacture of electrical generating plant (*i.e.* in the United Kingdom this carried the highest priority of all industrial processes).

Table 8. *Maximum Net Productive Capacity of Installed Electricity Generating Plant in E.C.O. area (in millions of kilowatts)*

	Productive capacity			Increase in productive capacity as compared with previous year		
	Thermal	Hydro	Total	Thermal	Hydro	Total
Total						
1938	25·0	14·0	39·0	1·0	0·5	1·5
1946	25·2	25·2	42·1	– – –	– – –	– – –
1947	26·3	26·3	44·0	1·1	0·8	1·9
1948	29·8	29·8	48·7	3·5	1·2	4·7
1949	33·7	33·7	54·2	3·9	1·6	5·5
1950	37·5	37·5	59·6	3·8	1·6	5·4
1951	40·9	40·9	65·5	3·4	2·5	5·9
Of which in contiguous countries†						
1938	15·0	10·6	25·6	0·4	0·4	0·8
1946	12·2	12·2	24·4	– – –	– – –	– – –
1947	13·2	12·9	26·1	1·0	0·7	1·7
1948	15·8	14·0	29·8	2·6	1·1	3·7
1949	18·2	15·2	33·4	2·4	1·2	3·6
1950	20·0	16·4	36·4	1·8	1·2	3·0
1951	20·7	18·3	39·0	0·7	1·9	2·6

* Participating countries and Western Germany.

† The countries are Austria, Belgium, Denmark, France, Italy, Luxembourg, Switzerland, and Western Germany.

Source: U.S. Department of State, Washington, Publication 2952: Committee of European Co-operation (European Series 29) published October, 1967.

Provision was also made for active international co-operation, *i.e.* certain stations built in one country but due to provide power for others, that is, Golderberg, Upper Inn, Weisweiler. The effect of these schemes is illustrated in diagram 1.

Conclusions

The objective of the 1948–51 plan was that by 1951 European coal production should have been restored to the pre-war level. In the interim assistance would be required in the form of imports on a gradually decreasing scale. These tonnages, which were exceptional in that Europe in the past had not been dependent on United States coal imports, were

Diagram 1. *Comparison of electricity production and requirements with and without the International Generating Plant Extension Programme*

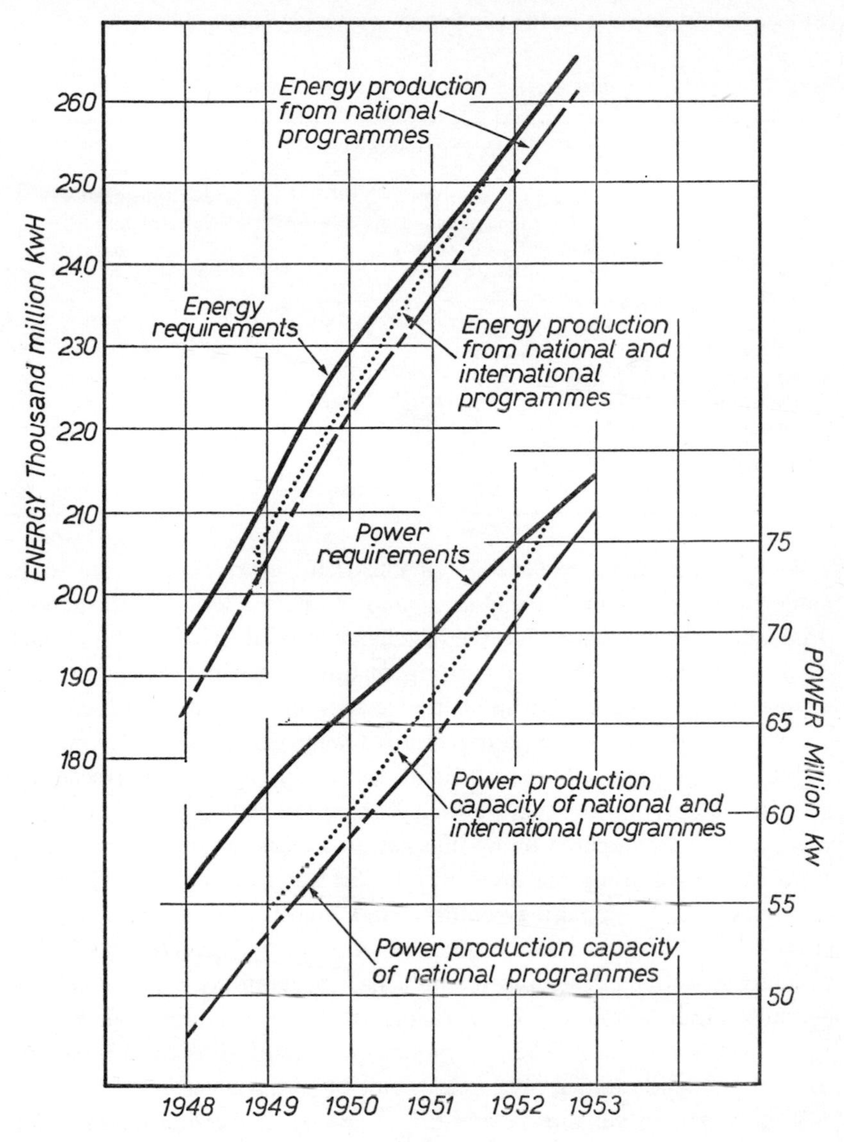

Source: U.S. Department of State, Washington, Publication 2952: Committee of European Co-operation (European Series 29) published October, 1947.

considered essential for the period under review to prevent the slowing down, if not the actual stoppage, of the 1947 pace of industrial recovery in all countries and to avoid the recurrence of the hardships endured by the peoples of Europe in the preceding winters.

Table 9. *Comparative energy consumption in E.C.O. area*

Source of Energy	1935–38 Average: Energy Consumption in terms of millions of tons of hard coal	1935–38 Average: %	1951: Energy Consumption in terms of millions of tons of hard coal	1951: %
Coal (including generation of electricity)	500	82	600	60
Water power (mainly for the generation of electricity)	50	8	100	10
Petroleum products and natural gas	60	10	300	30
Total	610	100	1,000	100

With regard to the generation of electricity great efforts were to be made in the participating countries and Western Germany to increase production and to make good the deficiency in productive capacity which grew up during the war years owing mainly to diversion of effort and resources. National generating plant extension programmes between three and four times as great as those of 1938 were undertaken. Despite this effort, it was expected that demand for electricity would exceed the available supply throughout the four-year period 1948–51. A programme was accordingly prepared on an international basis for further generating plant extensions using the most economical water power and brown coal resources. The participating countries and Western Germany undertook to make available the maximum contribution of new generating plant of which they were capable, and the industrial programme relied, therefore, on substantial supplies from abroad, notably the United States (i.e., equipment, know-how). The programme was designed to effect a balance between demand and supply of electricity by about 1952 or 1953.

With regard to oil, Europe had no possibility of meeting requirements from her own resources. The objective in this case for the period 1948–51 could be defined as one of limiting imports of oil to the strict minimum and to diversifying, wherever possible, the sources of supply in order to save on dollar expenditure.

The overall objective was therefore to seek to provide by 1951 total energy resources (i.e. coal, hydro power and oil) of about 1,000 million tons of coal equivalent (see Table 9 above).

Expressed in another way, the aim was to raise the annual consumption *per capita* from 2·3—which was the average for the period 1935–38—to 4·0 tons of coal by 1951.

The establishment of the European Coal and Steel Community

It was towards the end of the four year recovery plan that a new and potentially powerful factor was thrown into the European energy balance. This was the establishment in Luxembourg, under the presidency of M. Jean Monnet, of the European Coal and Steel Community, the forerunner of the much broader European Economic Community established five years later. On 9 May 1950, Mr Robert Schuman, then Foreign Minister of France, made his subsequently famous proposals for the pooling of the coal and steel industries of France and Germany, as well as any other European countries that might wish to join—including the United Kingdom. The French Government let it be known, however, that they considered that all countries that agreed to take part in the discussions with a view to establishing such a community should commit themselves, in advance, to the principle of placing their coal and steel industries under a common authority. It is sufficient at this stage to say that it seems overwhelmingly probable that the proposals were couched in this way because the French Government considered that by so doing they would make them unacceptable to H.M. Government—which although prepared to go far in the course of European co-operation, as it had already demonstrated in the O.E.E.C., could not at that time envisage yielding its sovereignty over two basic national industries.

The actual negotiations on the establishment of a common authority began in Paris on 20 June 1950, on the basis of a draft Treaty covering coal and steel, proposed by the French. The United Kingdom sent an observer. The negotiations soon ran into very severe difficulties over the problems of relationships between the new community and the Allied High Commission in Germany (which had the ultimate control over the German coal and steel industries). It seemed likely that under the provisions of the Schuman Plan vertical and horizontal concentration, and special sales agreements applying to the coal and steel industries, would be prohibited, or at the least rigorously controlled, in order to prevent the reconstitution of the big pre-war German coal and steel cartels, and the Germans accordingly sought to proceed and form a number of big mergers before the Schuman Plan came into operation. Another major source of difficulty during the negotiations was that the planning 'dirigiste' element among those responsible for, or favourable to, the effective operation of the Schuman Plan—and it was a very important one—which although appearing attractive in times of shortage, became considerably less so in the boom conditions engendered by the effects of the Korean War when producers could sell all the coal and steel they could provide. At the same time countries like Belgium and Italy, looking ahead to the time when the boom conditions then applying would come to an end, were anxious to protect their high cost coal and steel industries as much as possible and then provide them with temporary, short-term, safeguards, so as to avoid exposing them to the full effect of German competition.

By February 1951, the Schuman Plan negotiations were in danger of becoming completely bogged down, and it was only as a result of the strenuous efforts made by the Americans, who for major political considerations were determined to try and make the negotiations succeed, that a possible collapse was averted. Agreement was finally reached on 19 March 1951, when the leaders of the six national delegations concerned in the negotiations initialled the Treaty, establishing a Coal and Steel Community, on behalf of their respective Governments, and left only a number of political and constitutional points to be settled at a subsequent meeting of the Ministers for Foreign Affairs of the six countries. The Ministers for Foreign Affairs met in Paris in early April and the Treaty was finally signed on 18 April 1951. As a result of the hard bargaining that took place between Paris and Bonn, the Germans in return for their agreement to sign the Treaty of Paris extracted from the French an undertaking that they would put forward proposals for the removal of Allied controls on the German coal and steel industries. The Treaty represented a satisfactory bargain for both countries; for Germany it spelled the end of a period of irksome controls of her basic industries by the Allied High Commission and membership on equal terms with her five partners in the Coal and Steel Community. For France, it represented an indirect measure of control over the expansion of the German steel industry and access to German coke for her own expanding steel industry in Lorraine, and above all a first step towards the moral leadership of Western Europe to which she already aspired as the only means to re-establish her position in the Western world.

The members of the High Authority were nominated by the member Governments after the Treaty had been ratified by all six national parliaments (with the exception of the ninth member,[1] who was co-opted by his eight colleagues according to the provisions of the Treaty) and the High Authority officially opened its doors for the first time, in Luxembourg, on 10 August 1952.

Economically the Treaty of Paris was designed to create a single common market for coal, within which coal produced anywhere in the Community would be able to circulate freely. This was expected to bring about an overall expansion in coal output to meet the growing needs of an area that had a fast-growing and developing economy, albeit at the cost of reductions in output of coal in some of the older, exhausted, and therefore less economic coalfields. It represented, therefore, a deliberate effort on the part of the six member Governments to increase, and at the same time render more economic and less expensive to the consumer, the major internal source of energy in Western Europe. It was this first step, hesitant and uncertain, which was to lead, a number of years later, and only after the creation of the European and Euratom Communities by the Treaties of Rome in 1957, to the efforts to formulate a comprehensive common energy policy covering all forms of fuel and including a common policy on energy imports.

The Treaty of Paris, however, was designed as a much more modest—or pilot instrument. The fundamental aims and philosophy of the Treaty of Paris are set out in Articles 2 to 4 of the Treaty, and of these, Article 2 spells out better than any commentary the obligations which the European idealists set themselves when they hoped to shape the Treaty of Paris:

> 'The European Coal and Steel Community shall have as its task to contribute, in harmony with the general economy of Member States and through the creation of a common market in accordance with the provisions of Article 4, to economic expansion, growth of development, and a rising standard of living in Member States.
>
> The Community shall progressively bring into being conditions which will by themselves ensure the most rational distribution of production at the highest possible level of productivity, while safeguarding continuity of employment and being careful not to give rise to fundamental and persistent disturbances in the economies of Member States.'

The provisions of the Treaty can be divided into six main sections:

(i) the objectives of the Treaty (Articles 1 to 6);
(ii) the institutions of the Community and the division of power between them (Articles 7 to 45);
(iii) economic and social provisions (Articles 46 to 69);

[1] M. Paul Finet (Belgium).

(iv) transport (Article 70);
(v) commercial policy (Articles 71 to 75);
(vi) general provisions (Articles 76 to 100) and the Convention containing the transitional arrangements.

The Treaty, which was concluded for a period of fifty years, gave the Community a full juridical personality. Although based on the conception of a single market in coal and steel, it provided for a transitional period of 5 to 7 years in order to allow for its gradual progressive establishment and to avoid any possible serious economic or social disturbances. Its signature and acceptance by all six Governments committed them to the abolition of tariff barriers, subsidies and other special arrangements; and the elimination of discrimination in transport and general quantitative restrictions. For the coal and steel producers, it meant the compulsory abolition within the Community of discriminatory (and dual) pricing practices, restrictive practices, cartel agreements and associations, and unfair competition generally. Both Governments and producers remained free, however, to determine their own policies *vis-à-vis* third countries (although in steel the Community later moved towards a common external tariff).

One of the most vital and fundamental powers assigned to the High Authority was that embodied in Articles 58 and 59, which laid upon it the tasks of formulating and putting before the Council of Ministers of the E.C.S.C. proposals designed to deal with any severe surpluses and shortages of coal (or steel). Thus, instead of independent action being taken by the six Governments, it would have been the responsibility of the High Authority to determine the most suitable means of dealing with a given situation.[1] In the event, however, the coal crisis of 1958/59 demonstrated clearly that the vaunted supranational power of the High Authority was acceptable to the Member Governments only where it suited them and could easily be over-ridden when the measures proposed were unacceptable.

At the time of signature of the Treaty of Paris, costs and prices of coal in the six countries showed wide divergencies. There were no tariff

[1] Paragraph 1 of Article 58 stated that: 'In case of a decline in demand, if the High Authority considers that the Community is faced with a period of manifest crisis and that the means of action provided for in Article 57' (*i.e.* co-operation with Governments with regard to stabilisation measures; and intervention on prices and commercial policy) 'are not sufficient to deal with this situation, it shall, after consulting the Consultative Committee (*i.e.* a general advisory body in which producer, consumer, and trade union interests are represented) and receiving a confirmatory opinion from the Council, establish a system of production quotas . . .'

barriers (with the exception of a small tariff in Italy), but quantitative restrictions, or alternative methods of controlling imports, were in force in almost all countries (*i.e.*, Government-controlled import agency in France, price equalisation measures in force in several countries). Generally speaking, both costs and internal prices to the consumer were lowest in Germany, followed by Holland, France and Belgium, in ascending order. Therefore it was to meet this situation that the Transitional Provisions were cemented into the Treaty. Instrumental in this operation were the Belgians—who stood to lose most—and it was largely at their insistence that a perequation fund was set up, financed by a levy on the low cost German and Dutch coalfields (at a rate of 1½% of the proceeds, reducing over five years by 20% each time), which was to be used to assist the Belgian mines. Direct Government assistance in the form of subsidies was also authorised subject to the approval of the High Authority. It was furthermore agreed that Belgian coal production should not be permitted to fall by more than 3% annually in comparison with the output of the Community as a whole. There were also some special provisions to authorise the subsidisation of the high-cost Belgian production in special circumstances. Belgium, however, was not the only country to demand and obtain safeguards for her coal industry. France, although concerned mainly with her political objectives, obtained the inclusion of a provision that French coal production would not be allowed to fall by more than 1 million tons beyond a reduction proportionate to a general fall in coal output in the Community. To this end the High Authority was authorised to impose a levy of up to 10% on imports of coal exceeding the 1950 quantities from other Community countries. Special measures were also written into the Convention to help the small and very high-cost Italian coalmining industry. Although it is evident from the pains they went to in order to secure these safeguards that Belgium and Italy—and to a lesser extent France also—feared the competition from German coal, the general opinion in the Community was that demand for coal would continue to increase and to increase quite rapidly—but that most of this increase would come from German pits (and some also from the French Lorraine coalfield). At the same time, by 1952 the coal market situation was already very much easier than it had been in 1950 or 1951 when the Korean crisis had distorted normal market conditions. From the High Authority it was expected that it would seek to ensure that the coal (i.e., at this time the energy) requirements of the Community were adequately met; that the effect of shortages, where they occurred, would be distributed evenly throughout the Community and exports to third countries

—even if more profitable—should be restricted if and where this might prove necessary, in order to ensure supplies to consumers in the Community. The establishment of the Community was therefore expected to lead to some deflection of previously existing trade links and a substantial rise in the productivity, and hence relative commercial strength, of the Community's coal industries.

The Hartley Report

In view of the continuing rapid growth in the demand for energy in Western European countries and the consequent rise in prices and in imports, the Secretary-General of the Organisation for European Economic Cooperation in December, 1953, submitted to the Council of Ministers of O.E.E.C. a memorandum designed to draw the attention of all the Member countries to the growing problems of the supply and cost of energy and putting forward possible ways of overcoming them. As a result of this memorandum and subsequent discussions by the Council of Ministers, the latter, in January 1954, accepted the existence of an energy problem in Europe and called for additional information with regard to the existing situation. As a result it was decided to invite Mr Louis Armand (then Chairman of the French Railways and later the first President of Euratom) to undertake the preparation of a survey of all matters relating to energy which affected the Member countries of the organisation.

The report prepared by Mr Armand suggested that the era of fuel shortage which he described as the 'quantitative and technical' age was gradually being replaced by a new era in which the main sources of energy would be competing with one another. This new era, which Mr Armand described as 'productive and economic' would bring into the open the whole question of choice of the right type of fuel in the right market, that is, the principle of selectivity. Mr Armand had put on paper, publicly, and served notice that the era of fuel shortage was virtually at an end. He foresaw an era of competition between the different fuels. He did not and could not have foreseen the drop that was to occur in oil prices and the extent to which the shortage of fuel was to be replaced, after the interim period of the Suez crisis, by a vast energy surplus. Mr Armand based his case on the advent of atomic energy which, he suggested, had brought Europe to a turning-point. 'The history of energy resources is marked by sudden changes, veritable revolutions that constitute important stages in the advance of civilisation; the most characteristic of these stages was the introduction of the steam engine in the nineteenth century, followed, without any break in continuity, by the opening of the electrical and petroleum eras. The use of electricity and oil, which must be regarded as an evolutionary rather than as a revolutionary development, has not affected the history of the various forms of energy as much as the advent of the steam engine.' Whether this view is in fact correct is examined alsewhere in this book. ' . . . One of the first questions arising is whether the advent of atomic energy marks a turning-point similar to that of oil, and whether steps should not be taken now to ensure its rational development.'[1]

As a direct result of the Armand report the Council of O.E.E.C. in June 1955 passed two resolutions. The first called for the establishment of a Committee for Energy (and required it to submit a first detailed report to the Council not later than March 1956); the second concerned methods of co-operation in the peaceful use of nuclear power. It was the first of these that led to the publication of the Hartley Report entitled: 'Europe's growing needs of Energy: how can they be met?'

The Hartley report showed that between 1945 and 1956 energy consumption in all the Member countries had shown a steady rise and worked on the assumption that it would continue to do so. 'No doubt the repair of war damage and the making good of arrears of construction and equipment have influenced the high average rate of growth (4·8 %) since 1948. At the same time the demand for a higher standard of living and the need to increase productivity to enable Western Europe to maintain her economic position in the world are good grounds for assuming that this increase in the demand for energy will continue, although probably at a slowly decreasing rate. It is, in fact, essential, in order to preserve Western Europe's competitive position in world trade.' The report took 1948 as the basis year for all purposes of comparison, justifying this on the grounds that this was the year when conditions began to approach normality. Between 1948 and 1955 (although the latter figure given in the report was still an estimated one) inland consumption of primary energy in the Member States had risen from 526·6 million to 730 million tons (including the United Kingdom, where demand for primary energy rose from 220 million tons in 1940 to nearly 270 million tons by 1955). The figures for the individual fuels, expressed in tons of coal equivalent, are given overleaf in Table 10.

One significant trend emerged clearly from the report: the falling share of coal in the fuel market in relation to other fuels. Although the absolute

[1] *Some aspects of the European Energy Production.* Published by O.E.E.C., Paris, June 1955.

amount of coal required had increased during the period 1948–55 by almost 100 million tons, coal's share of the energy market fell from 78·9% in 1948 to 70·0% in 1955; almost the whole of this loss went to crude oil, which increased its share from 10·2% in 1948 to 17·2% in 1955.

Table 10. *Inland consumption of primary energy in the O.E.E.C. area (in millions of tons of coal equivalent)*

Year	Coal	Lignite	Hydro power	Natural gas	Crude oil	Total
1948	415·1	22·9	34·2	0·5	53·9	526·6
1949	440·4	25·1	33·1	0·8	61·7	561·1
1950	444·2	25·9	39·7	1·2	72·2	583·2
1951	494·6	28·6	45·5	1·6	83·7	654·0
1952	487·9	28·9	48·2	2·5	88·2	655·7
1953	478·1	29·2	49·0	3·6	98·8	658·7
1954	483·8	30·5	53·5	4·6	113·7	686·1
1955	511·0	31·0	57·0	5·0	126·0	730·0

Source: O.E.E.C: *Europe's growing needs of energy. How can they be met?* Paris, May 1956.

Out of the total demand for energy of 730 million tons of coal equivalent in the O.E.E.C. area, 584 million tons were met from internal production of coal, lignite, hydro-power and crude oil and natural gas. The balance of 146 million tons was imported in the form of coal and crude oil. The report emphasised, therefore, what was already known as Europe's energy gap. It was hard to believe that less than twenty years before, in 1929, Europe had still been an exporter of energy. Between 1948 and 1955 the European energy gap more than doubled, imposing a heavy strain on her monetary resources. The gap was met by importing coal, mainly from the United States, at an average annual rate of nearly 16 million tons, and oil where imports rose from 39 million to 84 million tons a year from 1948 to 1955.

Western European primary energy requirements in 1960 and 1975

In dealing with its main assignment, that is, a forecast of likely demand for energy in Western Europe over the following twenty years, the authors of the Report elected to take 1960 and 1975 as the two basic years: 1960 to give a short-term view in the belief that this would be relatively easy to determine accurately, and 1975 in order to give some guidance towards long-term investment. The Report's estimates were as follows:

	1955 (base year)	Millions tons of coal equivalent	
		1960	1975
maximum		860	1,300
mean	730	840	1,200
minimum		820	1,100

Diagram 2. *Primary Energy—demand and supply in the O.E.E.C. area (in million tons of coal equivalent)*

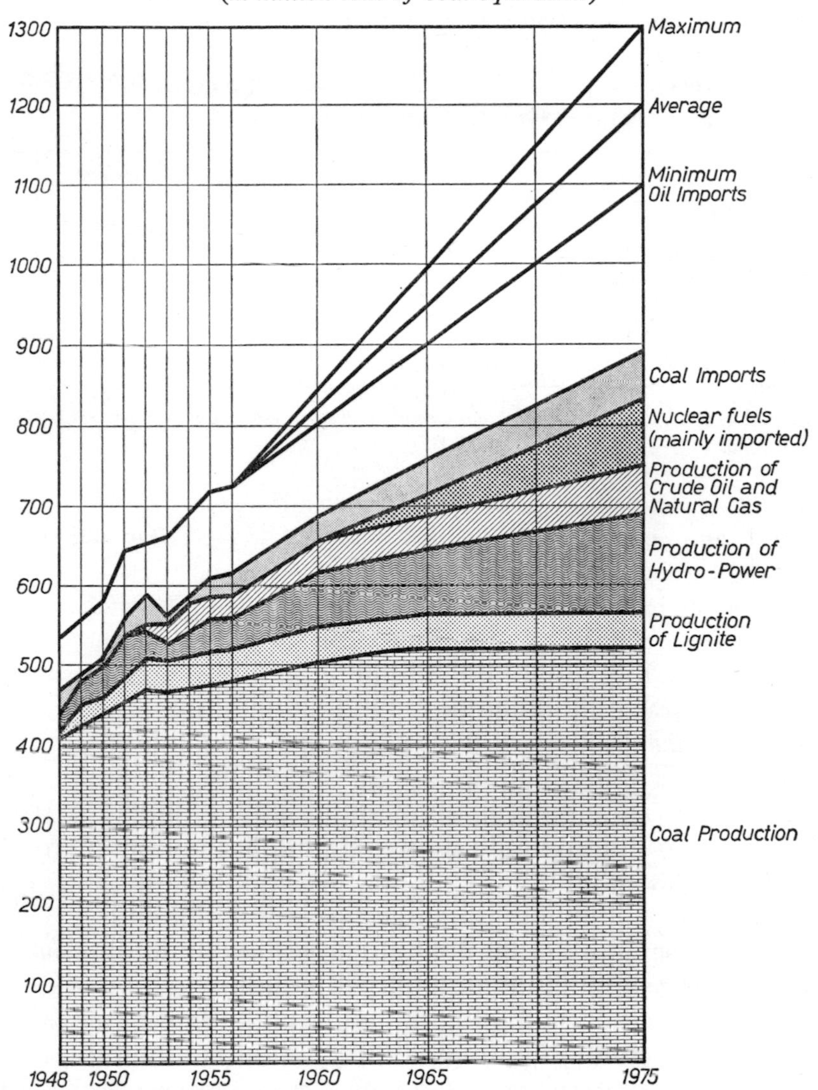

Source: O.E.E.C.: *Europe's growing needs of energy. How can they be met?* Paris, May 1956.

The share of each category of fuel for the twenty-year period 1955–75 forecast by the Hartley Committee is shown in the diagram on page 35.

The Report then turned to an examination of the likely output of energy from indigenous resources, and on the basis of the evidence and information provided by Government and industry representatives from the Member countries concluded that the level of output (expressed in million tons of coal equivalent) could be expected to rise from 584 million tons in 1955 to 645 million tons in 1975:

Table 11. *Indigenous production of all forms of energy in the O.E.E.C. area 1955–75 (in million tons of coal equivalent)*

	1955	1960	1975
Coal	478	500	520
Lignite	31	35	35
Hydro-power	57	75	130
Crude oil	13	25	50
Natural gas	5	10	20
Total	584	645	755

Source: O.E.E.C.: *Europe's growing needs of energy. How can they be met?* Paris, May 1956.

These estimates were based on the following assumptions:

Coal: an average rate of increased output of 1% per year until 1965, and the maintenance of the output in that year until 1975.

Lignite: an increased output of 2·5% per year until 1960, and a maintenance of output at the level of that year.

Hydro-power: an increase of 5·9% per year from 1955 to 1960 and 3·7% per year from 1960 to 1975.

Oil and natural gas: the output of both to increase fourfold over the period.

The main causes of the gap, the report continued, were simple to state. They were firstly the rapidly growing demand for oil—and Europe's lack of oilfields—and the slow rise in the output of coal when compared with the rate of increase in the demand for primary energy.

A simple comparison of the demand estimate and the forecast output of energy in the O.E.E.C. area demonstrated the full extent and seriousness of the energy gap; from 146 million tons in 1955, this was expected to grow to 195 million tons by 1960 and 445 million tons by 1975. Put in another way, the proportion of the total energy requirements which would have to be met by imports was expected to rise from 20% in 1955 to 23% in 1960 and to 37% by 1975. Thi increasing import requirement was regarded as constituting a double danger to the European economy. First, on the grounds that it would impose a severe strain on the European balance of payments; and secondly, that it would place Western Europe in a position of dependence upon extraneous sources. Calling energy the life-blood of Western Europe's industrial activity, the Report exhorted each Member State to develop home production to the fullest extent possible. In view of the fact that Europe had little or no resources of oil or natural gas and that economic hydro-electric resources would be used to the maximum extent by 1975, the Report concluded that coal resources should be developed as far as possible in order to try and reduce the gap.

Dealing with the individual sources of fuel, the Report considered that the contribution of nuclear energy towards meeting Europe's demand was unlikely to exceed 8% of total demand by 1975, or something like 85 million tons of coal equivalent. Imports of oil were expected to rise sharply, but there was no real evidence in 1955 that the world would be faced so soon with a surplus of crude oil and refined products. This was so true that many Governments were considering crash-programmes in order to hasten the advent of nuclear energy to supplement their limited internal supplies. Turning to coal, the Report stated that 'in the ample reserves of her coalfields, Western Europe possesses a great asset which not only can make a major contribution to her growing needs of energy, but is an essential part of the iron and steel industry, of the gas and coke industry and of many others. . . . It is not unfair to say that the industry as a whole has lost momentum which must be restored . . . It is essential in our opinion that the responsible authorities and the public should be made to realise the need for a new outlook on coal.'

'There is no doubt that given favourable conditions, coal could in time make a large contribution towards filling the energy gap. This will be dependent on price policy, on investment and on the recruitment of miners and engineers and scientists to aid in the development and more scientific organisation of the mines. A long-term policy is necessary with a stable outlook, favourable to investment, which will attract both young engineers and scientists who wish for a prospect of an interesting and attractive life's work in an industry which must have a great future as a vital part of Western Europe's economic structure. The future of coal must demand the closest attention of the Governments of all Member countries with reserves of coal and of the European Coal and Steel Community'. The Report accordingly concluded with a number of recommendations. After emphasising the fact that coal would continue to remain the mainstay of the energy economy in Western Europe for many years, the Report stated that an increased output of coal would depend on long-term investment, on the development of improved mining methods, on miners' pay and terms of employment being adequate and on giving coal the outlook of a modernised stable industry which will attract able young men to join it. In view, however, of the long-term commitments involved in investments in the coal industry, due account would have to be taken of possible changes in the competitive position of this industry once nuclear power has become a major factor in supply. The Report stated emphatically that all reserves of energy would be needed in the future. No doubt after 1975 nuclear power would free the coal burnt in thermal stations to an increasing extent, but this coal would be needed for other purposes where nuclear energy could not replace it.

In view of the probable limitations to the increase of productivity of indigenous energy in Western Europe, imports of oil would be needed in increasing quantities to bridge the energy gap. The world's reserves of oil were expected to be adequate to supply the crude oil required, but a very large investment would have to be made by the oil companies to enable them to meet the future Western European needs from the resources of their world organisation. The Report urged that competition would have to be allowed to play its full role in giving the consumer the maximum choice between energy sources. Nevertheless in order to deal effectively with the urgent problems involved in the supply and demand of energy, each Member country would require an energy policy suited to its own circumstances and its needs and resources. This policy should preferably include some measure of co-ordination between the different forms of energy.

The Report also called for due consideration to be given, when framing the overall economic, social and financial policies of each Member country, to the effects of taxation, price maxima and the wages structure on the efficient production and use of energy. Lastly, there should be increased co-operation between Member countries in the field of energy by means of exchanges of energy; by the supply of capital needed for the production of energy; by the exchange of manpower, information, and experience; by pooling research and development programmes; and by co-ordinating national energy policies.

No action in fact followed the publication of the Hartley Report, except that the O.E.E.C. Council of Ministers followed the recommendation to set up a permanent group of high-level experts to keep under continuous survey the problems of energy in Europe. The importance of the Hartley Report, however—as in the case of the Robinson Report published nearly four years later—lies in the insight it gives into thinking in Europe at that time on the whole question of energy supply and demand. The Hartley Report was quite correct in its assumption that the demand for energy would continue to grow—although it tended to underestimate the rapidity with which it would do so; but it proved grievously wrong about the need for Europe to expand, to the greatest possible degree, its own indigenous resources. Less than two years later Europe had moved into the 1958–59 coal crisis, the effects of which are still with us today. Oil flowed to Europe in abundant supplies and at prices well below the 1955 levels. Above all, the radical situation in Europe's balance of payments and the discovery of new oilfields in areas remote from the traditional sources of supply rendered meaningless almost overnight the arguments that the financial burden of such massive imports of energy would prove too much for the European economy or that excessive reliance for supplies on a given area of the world, prone to political instability, was strategically as well as economically unwise. It is, however, important to remember that when putting forward these arguments as recently as 1956, the Hartley Report was doing neither more nor less than reflecting the views of the vast majority of economists and experts in the field of energy at that time.

The challenge to coal

The conviction that the energy shortage, that had been a feature of the European scene from the end of the war to the middle 1950's, was to be a semi-permanent factor in energy calculations—a belief that had been reinforced by the findings of the Armand and Hartley reports and the directives given to the coal industries by all European Governments to maximise output—appeared to have been emphasised yet even more strongly by the effects of the Suez crisis which resulted in a comparatively short-lived but sensational four- to five-fold increase in Transatlantic freight rates. Such visible effects as the rationing of oil in the United Kingdom and the panic buying of American coal by German and Belgian importers, only provided additional manifestations of the generally-held belief that energy in the post-war world was a scarce and precious commodity. It was in this atmosphere that national Governments and experts in the field of energy continued to call for greater output of all types of fuel, while the E.C.E. Coal Committee in Geneva embarked upon an inventory of European energy resources[1] and the E.C.S.C. Council of Ministers first invited the High Authority (in 1957) to co-operate with the recently established European and Euratom Commissions in Brussels in formulating proposals for a co-ordinated energy policy for the six Common Market countries.

By 1957, however, although the first signs of the end of the era of fuel shortage were beginning to make themselves apparent, the time was not yet ripe for the full impact of this change to be realised. This indeed turned out to be a long and laborious process, often contested on both political and economic grounds. But one fact had become obvious. The supremacy of coal on the European energy scene, which in the immediate post-war years had been virtually complete, had been challenged, and with success, by oil. The rate of increase in the demand for energy in Europe had assumed such proportions that the coal industry, already straining to the utmost to increase its output, was unable to meet the requirements of the market. In these circumstances, oil stepped into the breach. In 1957, however,—and indeed for some years later—oil was supplementary to coal; it filled the gaps created by a soaring demand for energy and a relatively static or only slowly-rising production of coal. The beginning of the end of the fuel shortage had arrived simultaneously with the emergence of oil as a main source of supply. Together they heralded the end of the absolute rule of King Coal and the advent of a two-fuel economy.

In considering this period of flux we have followed the same pattern as in the earlier section dealing with the immediate post-war years, i.e. by taking each fuel in turn.

The fuel situation in the 1950s

Coal

By 1955, coal production in Europe had risen to 613·4 million tons (including 225·2 million tons in the United Kingdom). Production plans for 1960 showed a total of between 656 and 666 million tons (including 231·6 million tons for the United Kingdom).

In absolute figures, therefore, Europe was expected to increase coal production between 1955 and 1960 by only some 40 million tons. Even to achieve this modest target, the industry was faced with severe difficulties, of which lack of manpower was particularly acute. Between 1949 and 1955 manpower fell steadily in Belgium, France, and the Saar; in the

[1] In view of its size this table has been included in the statistical annex at the back of this book.

United Kingdom and Germany there were wide variations from one year to another, but the underlying trend was nevertheless a shift away from the pits. Only in Poland, among the major European coal-producing countries, had there been a steady rise in manpower. These fluctuations, and the very gradual rise in mechanisation of underground operations, resulted in a slow rate of increase in productivity in the mines, which in several countries was, in 1955, still below the pre-war level (i.e. Poland, Western Germany, and the United Kingdom). The only country to have shown any real improvement was France. At the same time, the average age of miners was rising steadily (thus, by the end of 1950, over 40% of the mine-workers in the Ruhr were over 40 years of age; in the United Kingdom the figure was just over 50%).

Table 12. *Hard Coal production in Europe in 1955, and forecasts for 1960 (in million metric tons)*

	1955	1960
Austria	0·2	0·2
Belgium	30·0	33·0
Bulgaria	0·3	1·4
Czechoslovakia	22·1	28·6
France	55·3	60·0
East Germany	2·7	2·9
West Germany (including the Saar)	149·1	158·8–163·8
Hungary	2·7	2·9
Italy	1·1	1·2
Netherlands	11·9	12·0
Poland	94·5	104–105
Roumania	0·2	0·2
Spain	12·4	15·6
Sweden	0·2	0·2
Others	4·9	6·9
Total	388·2	424·4–452·6
United Kingdom	225·2	231·6
U.S.S.R.	276·1	445·3

It is noteworthy that it was at about this time that the attitude of the consumers began to change and harden. The gradual improvement in the energy supply situation gave consumers for the first time since the war a greater element of choice, and this was reflected both in their attitude to coal as against competing sources of energy, notably oil, and with regard to coal imports both from other European and non-European sources.

Since the war the traditional European exporters had tended to regard their exports as a residual in their own balance sheets, so that in times of shortage the importing countries found themselves in difficulty and had to increase their purchases from extra-European supplies, that is, the United States. This generally resulted in a rise in the Transatlantic freight rates and a consequent increase in the landed price of American coal. The effect of this was to increase the pressure of demand on the cheaper European coal, which in turn pushed up the export prices of the latter coals. As soon as the peak demand subsided, freight rates dropped and American coal became cheaper than the European. Consumers then tended to prefer American coal until once again demand spurted forward. By 1954–58, however, consumers, believing themselves to be faced with a growing shortage of energy, began to show what was then a new interest in obtaining regular supplies of coal. Arrangements designed to secure this objective tended at that time, however, to run into the following obstacles: fluctuations in demand due on the one hand to factors beyond the control of consumers (climatic variations, changes in economic activity), and on the other to fluctuations in the use of alternative forms of energy; uncertainty of supply, especially during periods of shortage; instability of coal prices; and fluctuations in the qualities delivered. Any European coal policy, it was stated, should include as a major element the absorption of increasing quantities of European coal with as little fluctuation as possible and also take account of regular confrontation of long-range demand forecasts with producers' plans; a steadier flow of deliveries; arrangements to keep the price of coal in relation to that of alternative forms of energy; and the making of long-term trading arrangements.

Lignite

By 1955, consumption of lignite had come to account for about 20% of total European consumption of solid fuel. The importance of its role was enhanced by the fact that it was being produced in nineteen countries, and that production had risen from 288 million tons in 1950 to 401 million tons by 1955 (and was expected to reach 530 million tons by 1960).

In 1955, consumption of lignite and its derivatives accounted for more than two-thirds of the total consumption of solid fuel in Eastern Germany, Bulgaria, Hungary, Roumania, and Yugoslavia, and more than a third in Austria, Czechoslovakia and Greece. Together with Western Germany, these countries mined and consumed well over 25% of the lignite produced in Europe. Elsewhere, production of lignite was less important, either because of the small importance of the resources of lignite compared with those of other forms of primary energy, or as in Denmark and

Portugal, because of the unfavourable natural conditions for mining. Virtually, all the lignite produced was consumed within the borders of the producer country, and international trade in lignite accounted for less than 1 % of the total output; this is due to the exceptionally high moisture content of lignite—this can be as high as 60 %—which makes transportation of lignite in its crude state uneconomic.

Table 13. *Production of lignite in 1955 and forecasts for 1960 (in million metric tons)*

	1955	1960
Austria	6·6	7·0
Bulgaria	9·1	17·8
Czechoslovakia	40·7	63·0
East Germany	200·6	248·6
West Germany	90·4	121·0
Greece	0·9	2·7
Hungary	19·6	23·0
Poland	6·0	35·0
Roumania	5·9	11·0–11·6
Spain	1·8	3·0
Yugoslavia	14·1	26·9
Others	5·4	11·4
Total	401·1	530·0
U.S.S.R.	114·9	147·7

Petroleum

By 1955, proved European reserves were estimated to be (in ten million tons):

Western Europe	190	330
Eastern Europe	140	
Soviet Union	1,330	

The rate of annual increase in crude oil production from 1954 was 16 % for Western Europe and 7 % for Eastern Europe. Production in Eastern Europe was expected to increase from 12·6 million tons in 1955 to 14·2 million tons by 1960; in Western Europe an increase of 100 % in output was forecast.

Europe also had another source of production, based on coal. By treatment in coking and gasification plants coal could be made to yield benzene and heavy oils, and by treatment in hydrogenation plants various synthetic motor spirits; in fact, the entire range of petroleum products, with the exception of lubricants. Production from this source was expected to remain low in Western Europe where operators preferred to get more out of the coal by using the chemical by-products. By contrast, in Eastern Germany, Czechoslovakia, and Poland, production chiefly consisted of considerable quantities of liquid hydrocarbons made from coal, and especially from lignite.

Some crude oil was also produced from bituminous shales, but production from this source did not exceed 200,000 to 300,000 tons a year, and was not expected to show any great increase.

The consumption of crude oil in Western Europe increased by 90% between 1950 and 1955, thus marking the first real, massive breakthrough of oil into the European energy market. But there were substantial differences as between countries and products. Where these differences occurred between countries they were mainly due to the size of national energy resources and to certain amount of protectionism. Even so, crude oil consumption during this period had begun to rise even in such traditionally coal-consuming countries as Belgium, the Federal Republic of Germany, and the United Kingdom. Table 14 gives the consumption of petroleum as a percentage of total energy consumption of the countries of Western Europe in 1955 (comparable figures for Eastern European countries are not available).

Table 14. *Relation of petroleum consumption to total energy consumption in Western Europe in 1955*

Country	Petroleum consumption as % of total energy	Country	Petroleum consumption as % of total energy
Western Germany	9	Switzerland	25
United Kingdom	12	Italy	32
Austria	15	Denmark	38
Belgium/Luxembourg	16	Sweden	43
France	22	Portugal	45
Netherlands	25	Greece	70
Norway	25		

Consumption also varied with the product. Thus, the consumption of motor spirit as a percentage of that of all petroleum products fell from 25% in 1950 to 21% in 1955. The proportion of paraffin also fell, from 7% to 5%, while that of gas oil remained steady at about 22%. The consumption of fuel oil, which had accounted for only 35% of all petroleum products, rose to 40%.

Petroleum products had to compete on the European market with the other sources of energy that were available. However, the greater part of such products used for generating motive power—89%—was absorbed by transport, the cardinal use to which these products were put. In addition the growing proportion of diesel-engined road-transport vehicles was also pushing up the consumption of gas-oil.

The consumption of petroleum products for generating heat also increased considerably during this period, despite the competition from other sources of energy in this field. Thus, in the power stations, petroleum products accounted for one-fifth of the increase in output between 1950 and 1955, although solid fuels continued to meet the main needs of these plants. At the same time the consumption of black oils in the domestic market increased considerably, although here again coal continued to play the leading part as it was still used almost exclusively for space heating in the Federal Republic of Germany and in the United Kingdom. It is also noteworthy that for furnace use in the metal industry, the consumption of coal and other solid mineral fuels increased only very slightly, whereas the consumption of petroleum products quadrupled over the period 1950–55. As a result, consumption of black oils in Western Europe in the power stations and in the domestic, and iron and steel markets had, by 1955, risen to more than 8 million tons, more than half of which was gas-oil.

The breakdown of petroleum consumption in Western Europe by 1955 is shown below:

Table 15. *Breakdown of petroleum consumption in Western Europe*

	1955	
	in million tons	%
Motor spirit (including white spirits and aviation fuel)	23·4	23
Intermediate fractions (paraffin, jet fuels)		
Internal consumption	24·4	24
Bunkers	3·2	3
Fuel Oil		
Internal consumption	32·3	31
Bunkers	8·8	9
Other products (lubricants, bitumens, etc.)	10·9	10
	103·0	100

In Western Europe, while industrial production increased by approximately 8% a year between 1945 and 1956, refinery capacity increased from 15 million to nearly 120 million tons. As a result of this development the following plant capacities had been reached at the beginning of 1955 and 1957 respectively (in millions of tons a year):

Type of Plant	1955	1957
Distillation	126·705	134·262
Thermal cracking	5·391	5·571
Catalytic cracking	19·400	23·907
Thermal reforming	8·928	8·242
Catalytic reforming	3·992	7·852
Polymerization	1·239	1·352
Hydrogenation	1·411	1·373
Lubricants	2·790	2·903

It will be noted that catalytic reforming plant capacity increased considerably, even over this comparatively short period, whereas catalytic cracking was formerly the most important process. This trend, which continued subsequently, helped to adjust the pattern of refinery production to the range of demand in Western Europe. Whereas up to 1955/56 the development of refinery capacity was more or less geared to the quantity of crude oil to be processed, the same was not true of the range of products. The main reason for this anomalous situation was the practice of fixing world prices of petroleum products on the basis of the North-American market, where the consumption pattern was very different from that in Western Europe:

Range of petroleum products consumed

	Western Europe	U.S.A.
	%	%
Motor spirit	21	42
Paraffin	4	4
Gas diesel oils	23	20
Fuel oil	40	20
Other products	12	14

Diesel oil was being used more and more for driving transport vehicles and agricultural machinery, whereas industry preferred fuel-oil; light oils were being used more and more for space heating. As a result, the demand for intermediate fractions and fuel oils was increasing rapidly while the consumption of motor spirit tended to remain steady, or at best to increase more slowly with the emphasis on the higher qualities. Furthermore, there was no agreement between the refinery operators and the motor-engine manufacturers as to whether the engines should be designed for high or low octane motor-spirit, or in the case of diesel engines for high or low cetane fuels.

Theoretically, it seemed possible to change the European refineries' programme;[1] the feasibility of such a change depended exclusively upon its cost, which was then estimated by one report at £1,200 million. This would have entailed an additional charge of U.S. $0.40 per barrel of crude oil treated, that is, 20% of the then current price f.o.b. Persian Gulf. Such a charge would have made it possible to double the output of fuel oil while increasing that of motor spirit by only 50%.

Unless there was a corresponding drop in the price of Middle East crude, the adaptation of refinery programmes to European consumption was expected to lead to higher prices for fuel oils and the intermediate fractions rather than for motor spirit, which was already considered to be too high as a result of the tax it had to carry. In 1957 it was believed in some quarters that such a movement had already set in.

In Eastern Europe, refinery capacity at 1 July 1956, was reputed to be as follows[2] (in million tons a year):

Country	Number of Refineries	Crude Oil Capacity	Cracking Capacity
Albania	1	0·2	
Bulgaria	1	0·45	
Czechoslovakia	1	0·43	
Hungary	3	1·1	
Poland	5	0·5	
Roumania	—	9·6	
Yougoslavia	3	1·0	0·25
U.S.S.R.	70	77·2	32·0

The refineries in this part of Europe have had to be operated at full capacity since the end of the war so as to treat the increasing quantities of crude oil that had become available.

Information with regard to Eastern European countries is difficult to obtain. Available information on oil refining in Poland suggests that by the end of 1955 refining capacity there had risen to 500,000 tons—although actual output of finished products was only about 400,000 tons. This seemed to indicate a serious loss, and it was therefore planned to build a new refinery with a capacity of one million tons by 1960. This was designed to reduce production costs and to enable Poland to produce better finished products from the varied range of crude available. The five refineries operating in 1955 satisfied only 40% of current internal requirements.

In the Soviet Union refinery construction during the post-war period, although meeting the targets set out in the 1950–55 five-year plan, did

[1] See *The Institute of Petroleum Review*, London, October 1956.

[2] See *World Petroleum*, July 1956.

not keep pace with actual crude-oil production, thus suggesting that the Soviet Union intended to export some qualities of crude. In 1938, the Soviet Union, for an output of 30·2 million tons, had 28·4 million tons of ordinary refining capacity and 9·3 million tons of cracking capacity. By 1942, these capacities had been increased by about 10%. Between 1950 and 1955, as a result of reconstruction and development plans, there was a very rapid increase both in production of crude oil and in refinery capacity: (in million tons)

	1950	1953	1955
Crude-oil production	37·9	52·8	70·8
Ordinary refining capacity	36·3	47·0	65·0
Cracking capacity	13·5	15·0	32·0

An important feature of the situation was that the refining industry was gradually being moved away from European Russia and towards the centre of the country. Thus, whereas in 1955 the U.S.S.R. had some 70 refineries in all, only a few of which were situated in the eastern region of Russia, in the beginning of 1957 a large refinery was put into operation at Omsk in Western Siberia, and other refineries were in the process of construction at Krasnoryarsk in Central Siberia and Amur-Oblast in the Pacific area. It was intended that by 1960 the capacity of the Siberian refineries would be considerably greater than that of the refineries in the Baku region, the more so since other refining plants outside the European part of Russia were then being planned at Uzbek and Kazakhastan, both in Central Asia. This shift in petroleum processing from the West to the East of the U.S.S.R. was no doubt part of the Government's deliberate policy of establishing new industrial centres in the eastern part of the country.

Natural Gas

In 1957 the development of natural gas in Europe was still limited to a small group of countries: Austria, France, Italy, Poland, Roumania, Yugoslavia, and the Soviet Union. The characteristics of natural gas (methane) are such as to justify its comparison with hydro-electric power or petroleum. Like the latter, it has a high calorific value (8,500 to 9,500 kcal/m^3), and although not lending itself to such universal use, it is a practically pure fuel and it does not need to be stocked on the consumer's premises. For many uses it is, in fact, better than fuel oil (*i.e.* in the United States, for example, the advantages are considered to justify a price per calorie 10% higher than that of fuel oil).

Natural gas has substantial advantages over manufactured gas. In the first place it is much cheaper to produce. In France, for example, where the production costs of Lacq gas were, and still are, burdened by certain charges due to the presence of large quantities of hydrogen sulphide and to the depth of the deposit, it was still possible to sell the gas at a distance of 300 kilometres from its point of origin at a price per calorie 50 % lower than that of Lorraine gas at the coke oven plant. This was because, apart from its production cost at the well, transport was less costly for natural gas since its calorific value per m^3 was more than twice that of manufactured gas. It follows that a pipe of a given cross-sectional area carried a much larger number of calories of methane than of manufactured gas, as the weight per unit length of the pipe varies according to the square of its outside diameter. Accordingly, natural gas is ideally suited for long-distance transport, as practised in the United States and in the U.S.S.R.

The features of natural gas were and are such that the demand for it in Europe is practically unlimited and its development will be restricted only by the supply available.

In 1957 Europe's resources of natural gas were estimated to be as follows (in thousand million m^3):

France	300
Italy	130
Western Germany	35
Austria	18
Yugoslavia	18
Total for Western Europe	501
Roumania	246
Hungary	4
Poland	4
Czechoslovakia	3
Total for Eastern Europe	257
Soviet Union	985

Hydro-Power

Since 1938 the share of hydro-power in total electrical power consumption remained at about 37 % for Europe (excluding U.S.S.R.) and 13 % for the Soviet Union. In 1955 its share in the overall energy balance was 6 % for Europe (excluding U.S.S.R.) and 2 % in the Soviet Union. The average annual rate of increase of hydro-power between 1950 and 1955 was 8 %, as against 5·3 % for primary energy as a whole.

A study by the E.C.E. Committee on Electric Power, published in 1958, drew attention to development possibilities in the countries of Eastern and South-Eastern Europe. These mainly concerned the Aliakmon (Greece),

Lakes Prespa and Ohrid (Albania, Greece and Yugoslavia), the Danube and the 'Iron Gates' (Roumania and Yugoslavia), and the Vistula (Poland).

Table 16. *Degree of Exploitation of Hydro-Electric Power Resources, 31 December 1956*

Country	Maximum exploitable hydro potential: Total per inhabitant (kWh/year)	Percentage harnessed at end of 1956	Remaining to be harnessed (in million kWh/year)	Hydro production in 1956 as % of total electric power output
Albania	3,587	1	4,950	37
Austria	5,736	22	31,350	74
Belgium	61	34	350	2
Bulgaria	1,470	7	10,350	32
Czechoslovakia	917	16	10,100	11
Denmark	11	60	20	—
France	1,762	34	50,250	48
Finland	4,008	30	11,850	76
East Germany	111	23	1,550	2
West Germany	525	151	12,420	16
Greece	534	12	3,700	33
Hungary	342	2	3,300	1
Italy	1,145	57	23,700	77
Luxembourg	210	6	50	—
Netherlands	—	—	—	—
Norway	30,511	23	80,950	99
Poland	488	5	12,650	4
Portugal	1,506	15	11,160	94
Roumania	1,551	1	26,700	6
Sweden	11,016	31	55,550	90
Switzerland	6,490	45	17,650	98
Others	—	—	124,800	—
United Kingdom	225	20	9,210	2
Total	1,865	23	586,400	30
U.S.S.R.	7,490	1	1,471,000	16

Source: United Nations Statistics: 1957 and 1958.

Secondary Forms of Energy

Electric Power. Between 1946 and 1951 the average annual rate of growth in consumption of electric power in Europe was 11·5%. Between 1951 and 1956 there was a slight fall to 9·5%. This was an overall rate, which therefore far exceeded the level of 7·3% which corresponds to the law of doubling every ten years. The annual rate of increase in output exceeded 10% in Austria, Finland, Greece, Hungary, Iceland, Poland, Portugal, Roumania, the Saar, Spain, Turkey, and Yugoslavia (also in U.S.S.R.); it was between 7·5 and 10% in Czechoslovakia, Denmark, Eastern and Western Germany, and the Netherlands. The following two tables show, first, the development of electric power consumption in industrial, domestic, and transport categories, from 1946 to 1956; and secondly, the distribution of total electric consumption in Europe (including U.S.S.R.) in 1955 over the three main sectors:

Table 17. *Development of Electric Power Consumption in Industrial, Domestic and Transport Categories, 1946–56* (1946: 100)

	1951			1956		
Country	Industry	Domestic	Transport	Industry	Domestic	Transport
Austria	337	157	188	526	290	262
Belgium	156	127	201	195	171	188
Bulgaria	275	167	133	550	457	542
Finland	161	137	117	242	275	158
France	180	138	124	250	240	201
Germany W.*	253	133	195	420	241	287
Italy	183	132	193	251	209	246
Netherlands	225	213	319	243	321	542
Norway*	168	145	135	223	216	187
Sweden	138	166	121	196	248	143
Switzerland	133	126	117	164	188	137
United Kingdom	142	151	111	191	222	100

* refers to 1947 as base year.
Source: United Nations Statistics: 1951–56.

Table 18

	In thousand million kWh	As percentage of total electric power consumption
Industry	398	72·8
Transport	24	4·4
Other consumers (domestic, handicrafts, agriculture, commerce, etc.)	125	22·8
Total net consumption	547	100·0

The period of intense economic activity—which characterised the greater part of Europe throughout this period—had of course, as Table 17 shows, the effect of boosting the consumption of electric power both in industry and in the domestic sector. It was significant, however, that their increase was substantially over and above that due to the steady development of electrification and of the uses of electricity. This phenomenon was expressed in the following formula by Monsieur P. Ailleret:[1]

$$\text{electrical power consumption} = (\text{index of industrial activity})^n \times \text{the time exponential.}$$

At that time the price of electricity in a number of European countries had not followed the trend of the prices of other forms of energy, nor, more generally, of the cost of living, and the balance between revenue and expenditure could very often be kept only by adopting low depreciation rates. This sometimes led to the substitution of electric power for other forms of energy—a development which was not always economically desirable and did not offer sufficient incentive to economy.

The net production of thermal power stations in European countries as a whole in 1955 (including the United Kingdom) was 406,000 million kWh, 70% of their electric power production. This figure rose by 3% each year between 1951 and 1955. Construction costs decreased during the period, mainly as a result of the practice of using larger units. At the same time the efforts to increase the efficiency of thermal power stations by working at higher temperatures and pressures involved using more costly materials. The average cost of a modern thermal coal-burning power station in 1955 was estimated at £51·5 per kW (although the cost of a plant burning brown coal was estimated at only £30·0 per kW). Together with this, however, was an increase in firing costs. These were primarily determined by specific consumption, which in most modern European power stations amounted at that time up to 2,500 to 2,800 kcal/kWh.

It was generally believed in 1955/56 that in the case of thermal power stations using good quality coals or fuel oil, the decline in fixed charges and in specific consumption would not be sufficient to offset the anticipated increases in the price of fuel, but that the cost per kWh produced in thermal power stations burning low grade fuel would continue to fall.

The following table shows the trends in the consumption of the various fuels in power stations from 1952 to 1956:

[1] Economic Studies, International Union of Producers and Distributors of Electric Power (UNIPEDE), London Congress, 1955.

Table 19. *Annual consumption of fuels for electric power production in fifteen European countries taken as a whole* (in thousand million kcal)*

	Coal: Colorific value more than 4,500 kcal/kg	Coal: Calorific value less than 4,500 kcal/kg	Lignite, peat, or wood	Liquid fuel	Natural and manufactured gas	Other fuels	Total consumption
1952	587,977	128,363	90,012	45,089	28,166	856	880,463
1953	617,903	140,525	98,292	50,406	32,560	856	940,543
1954	668,684	153,219	105,241	59,764	34,209	859	1,021,975
1955	702,894	176,159	116,350	63,358	46,682	983	1,106,426
1956	745,510	199,832	128,498	79,663	58,164	1,217	1,212,884
1956 as percentage of total consumption	61·5	16·5	10·6	6·5	4·8	0·1	100
Increase from 1952 to 1956 (4 years) (as %)	27	56	43	76	103	42	38

* The fifteen countries are: Austria, Federal Republic of Germany (public service only), France, Greece, Hungary, Ireland, Italy, Luxembourg, Netherlands, Norway, Poland (public services only), the Saar, U.S.S.R. (public services only), United Kingdom (in the case of liquid fuel, public services only), and Yugoslavia.
Source: United Nations Statistics: 1952–56.

It will be seen that in the countries considered, coal consumption rose by 27% during the period 1952–56, whereas fuel consumption as a whole increased by 38%.

One of the most notable post-war developments was the trend towards electrification of railway lines. Figures illustrating this are shown in Table 20.

In view of the exceptional importance of power-generation in the European energy scene, tables showing: (*a*) composition of inland consumption in Europe of primary energy from commercial sources in 1955; (*b*) the main sectors of electric power consumption by industry in 1956; (*c*) development of electric power production in Europe from the end of 1956; have been included in the appendix at the back of this book.

Gas industry and coke ovens. Leaving natural gas out of account (and in 1956/57 it was still a minor contribution to Europe's energy requirements) the production of manufactured gas in Europe was, in 1956–57, still intimately bound up with iron and steel production. This was the

Table 20. *Electrification of European railway systems, 1946–56*

Country	Total length of line operated: 1946, In thousand km	1946, Percentage electrified	1956, In thousand km	1956, Percentage electrified
Austria	5·80	17·8	5·99	27·7
Belgium	4·96	0·9	4·95	15·4
Czechoslovakia	13·10	0·6	13·17	2·1
Denmark	2·41	1·7	2·56	2·3
Finland	4·43	—	5·00	—
France	40·63	8·6	39·83	13·9
Greece	0·18	—	1·74	—
Hungary	7·93	1·8	8·00	—
Italy	14·78	25·2	17·08	37·4
Luxembourg	0·54	—	0·39	9·2
Netherlands	2·97	6·0	3·22	43·8
Norway	4·28	15·9	4·42	29·7
Poland	24·39	—	—	—
Portugal	2·52	—	3·59	0·8
Roumania	10·25	—	—	—
Spain	12·80	5·2	13·01	11·4
Sweden	12·55	37·3	14·93	43·7
Switzerland	3·21	90·7	2·93	97·8
Western Germany	—	—	30·56	7·1
Yugoslavia	—	—	11·74	1·4
United Kingdom	31·96	—	30·65	5·3
U.S.S.R.	—	—	120·70	4·5

Source: United Nations statistics.

culmination of a process which was nearing a major cross-roads. The production figures for manufactured gas in certain European countries for 1955 were:

Country	Gas distribution in 1955 (in million m³)
France	4,653
Netherlands	2,267
Western Germany	12,244
Poland	2,456
Czechoslovakia	1,247
	22,867
U.S.S.R.	13,410
United Kingdom	20,410
	33,820

Development of non-conventional resources in Europe

Much consideration was being given in the middle of the 1950s to the development of non-conventional sources of energy. These were as follows: the tides; artificial rain; the swell of the sea; the thermal energy of the seas; solar energy and geothermic energy. The only one of these likely to be of any commercial importance in the European area is energy tapped from the ebb and flow of the tides. The energy stored up in the tides is produced by the earth's rotation on its own axis in the luni-solar gravitational field. But the tides cannot be used economically unless they are of sufficient amplitude, which occurs rarely in Europe; in fact, only in the Severn Estuary in England, and the Mont St Michel Bay in France. In 1957 there were three big projects under examination. The first of these, known as the Rance project (340,000 kW), was already in course of execution. The second, still in a preliminary planning stage, was the Chausey Islands project, which was due to have a capacity of some 10–15 million kW. In England, the Severn project—which was later abandoned—provided for a capacity of 800,000 kW, with a production of 2,300 million kWh. There was so little correlation between the irregularity of the tides and the irregularity of demand that if the tidal power stations had been equipped with conventional turbines, the energy produced would not have been competitive in cost with conventional energy. But the electricity networks had become so interconnected that they were beginning to be able to absorb large irregular supplies of power. Moreover, optimum utilisation that would both save fuel and boost supplies at peak hours could be obtained with the then recently developed bulb-type units, which were equipped with turbo-generators that could turbine the water in both directions, and also pump.

Imports

In 1955, petroleum made up approximately 80% of Western Europe's energy imports—coal accounting for the rest. But whereas coal imports varied with economic conditions and as a result of the lack of flexibility in European coal production, petroleum imports showed a steadier increase, corresponding to that in total energy requirements. One result of this was the development of bigger tankers and the construction of oil pipelines. By the end of 1956 the pipelines shown in Table 21 were in operation.

On the coal side, imports from the United States rose from 25 million tons in 1955 to 39 million tons in 1956, and over 45 million tons in 1957. Until then U.S. coal had been regarded as a temporary stopgap, to which

Europe only had recourse on a large scale during emergencies, such as the Korean War in 1951 and the Suez crisis in 1956.

Table 21. *Pipeline specifications (1956)*

Pipelines operating	Length (miles)	Diameter (inches)	Approximate Capacity in millions of tons per year	Pumping Stations	Year completed	Estimated cost in million U.S. dollars
TAP line	1,068	30–31	15	6	1950	230
Kurkirk-Damascus	555	30–32	14	7	1952	115
Inter-provincial	1,126	16–20	4	7	1950	90
Trans-mountain	718	24	6	3	1953	92

Natural gas

The need to free the gas industry from too close a dependence on coke-oven plants, whose level of operation depended on circumstances often unrelated to those governing the demand for gas, had already, for some time, fixed attention on the prospects of importing the unused natural gas available in substantial quantities outside Europe. Among the various projects considered, the Bechtel scheme, which was intensively studied between 1951 and 1955, envisaged the conveyance of natural gas from the Middle East to Europe by pipeline. The pipeline was to be 4,000 kms long, and was to run from Iraq through Turkey, Greece, Yugoslavia, Austria, and South Germany, to Northern France. Two stages were contemplated, under which one pipeline equipped with the requisite pumping stations was to be built first, and then duplicated later by a second. The volume of gas transported would have been 9,600 million m^3 a year. The capital outlay required was estimated at the time to amount to £277 million, of which £152 million would have been needed for the first stage, and £125 million for the second.

In 1951 the average cost of the gas conveyed was estimated at 11·14*d.* per thousand Kcal, a figure which was lower than the cost estimates for liquefied natural gas imported from Venezuela. This figure, however, represented the mean cost, and since the actual cost would vary according to the distance travelled, it would have been higher at the end of the pipeline (i.e. in Germany and France). It was subsequently suggested that a project of this kind might be of more interest to the less distant countries, such as Greece, Yugoslavia, and Hungary, and perhaps Austria and Czechoslovakia. In particular, it might usefully have been combined with a scheme for exports from Roumania.

The importing of natural gas liquefied at low temperature (−160°)[1] in specially designed tankers was also carefully examined at about this time. According to one working hypothesis at this time the price of liquefied natural gas delivered at destination would have been well under 50%—subject to fluctuations from one region to another—equivalent to half the price for manufactured gas:

Table 22. *Range of liquefied natural gas prices in relation to gas prices in a number of European countries*

Countries of Origin	Destination	Currency Unit	Cost of liquefied natural gas at place of destination	Present price of gas at place of destination	Ratio columns 4/5 %
Middle East	London	Pence per 1,000 cu. ft. at 4,200 Kcal	25	46	54
Middle East	European centres	Francs per m³ at 4,200 Kcal	5·30	industry 13·50 average 22·95	39 23
Middle East	London	Belgian Francs per m³ at 4,200 Kcal	0·49	0·90	54
Saudi Arabia	U.K.	—	—	—	50
Caribbean	London	Pence/therm	5·5–6·0	industry 13·32 average 17·87	43 32
Venezuela	Le Havre	Centimes/therm	0·90	industry 3·17 average 5·76	28 16

As this table shows, for the same calorific value, the cost of natural gas delivered to Europe would not in any case have exceeded one-half of the cost of manufactured gas delivered in Europe. It was estimated, moreover, that in London it would amount to approximately 70% of the cost of its equivalent in fuel oil. On the other hand, in the southern half of France it would have been slightly more expensive than Lacq gas, and in Italy more expensive than the natural gas from the Po Valley. In France the discovery of the Lacq natural gas deposits in the early 1950s had led to a substantial pipeline system in France. The real expansion of oil and natural gas pipelines in Europe did not come about, however, until the end of the 1950s and the early 1960s.

[1] At this temperature the gas occupies about 1/600 of its gaseous-stated volume.

SECTION 2

THE YEARS OF SURPLUS FROM 1957 TO THE PRESENT DAY

The growing importance of oil

After the rapid rise of overall energy consumption that had characterised the first post-war decade and which continued up till 1956, there followed a short two-year period, in 1957 and 1958, during which the rate of increase in general industrial activity in Europe levelled out and demand for energy showed little change. By 1959, however, the upward trend both in industrial activity and overall energy consumption had been resumed. Despite this temporary slackening in the overall demand for fuel, oil consumption between 1955 and 1959 in the countries of Europe participating in the O.E.E.C. increased by some 50 million tons: oil consumption more than doubled during this period in West Germany, while in Italy, the Netherlands and Switzerland there were increases of over 60 %; by the end of 1959 oil's share of the energy market in O.E.E.C. countries had risen to 30 %, compared with only 21 % four years earlier. Within this trend there were wide variations in the relative importance of oil in individual countries: in Sweden, oil's share rose from 55 % in 1955 to 64 % by 1959, and in Italy, from 44 % to 52 %. Neither of these two countries had an indigenous coal industry of any size and both relied to a considerable extent on hydro-electricity. Oil consumption rose more slowly in the big coal-producing countries, but even here there was a steady tendency to move towards oil, illustrated by increases in oil's share of the market over the 1955–59 period from 27 to 31 % in France, from 19 to 30 % in Belgium, from 10 to 20 % in Western Germany, and from 15 to 24 % in the United Kingdom.

From 1960 to 1964 there was a further rise in oil consumption in the European area of the O.E.C.D.[1] of nearly 70 %, *i.e.* from 181 to 306 million tons. There was also, during this period, an increasing trend in Western Europe towards using gas/diesel oils as a result of the growing use of these oils for smaller domestic heating appliances and the rapidly

[1] The former O.E.E.C. was re-organised in 1960 and became the Organisation for Economic Co-operation and Development (O.E.C.D.), with the United States and Canada, and later Japan also, as full members.

rising number of diesel-engined heavy road vehicles. The largest increase, however, occurred in the consumption of fuel oil, which rose by nearly 80 %.

Table 23.[1] *Oil products consumption pattern in O.E.C.D. European area, 1960–64 (in million metric tons)*

Year	Refinery Gas	Liquefied Gas	Aviation fuels, Kerosine and other products	Motor Gasoline	Gas Oil and Fuel Oil	Total
1960	1·0	3·2	9·5	30·7	111·9	156·3
1961	1·0	3·6	10·9	33·6	129·6	178·7
1962	1·3	4·2	11·8	37·0	148·0	202·3
1963	1·6	5·3	12·1	39·9	167·5	226·4
1964	1·9	6·3	12·9	43·2	196·8	261·1

In its 1964 report 'Oil Today' the O.E.C.D. Special Committee for Oil found that the greatest rate of growth in oil consumption was in public electricity generation and gas-making—where there had been increases in consumption between 1955 and 1959 of some 68 %. But high though this seemed, the rate of increase in the demand for oil was to accelerate even more rapidly between 1960 and 1964 to reach a figure of some 150 %. Since such a rate of growth far exceeded the general rate of growth in industrial markets, this meant of course that petroleum products were being used proportionally more in these markets at the expense of other sources of energy. Big increases in the use of oils were also reported in the domestic and general industrial markets. As a result oil's share of the overall energy market increased further between 1959 and 1962 from 30 to 39 %, and by 1965 was only fractionally short of 45 %. This advance was made almost exclusively—as had already largely been the case during the period from 1955 to 1958—at the expense of coal:

Table 24. *Pattern of O.E.E.C./O.E.C.D. primary energy consumption 1955–64 (in %)*

Year	Solid Fuels	Petroleum Products	Natural Gas	Hydro Electricity	Total
1955	75	21	1	3	100
1957	74	22	1	3	100
1959	64	30	2	4	100
1962	55	39	2	4	100
1963	53	41	2	4	100
1964	49	45	2	4	100

[1] See *Basic statistics in energy: 1950–64*, published in 1965 in Paris by O.E.C.D.

The main reason for this astonishingly rapid increase in the use of oil was undoubtedly the greater convenience of liquid fuels. Until 1957 oil was still regarded as an essential supplementary source of fuel which was required to augment Europe's indigenous supplies of energy. At the same time there were of course certain uses, such as road, water and air transport for which there was no alternative to oil and these sectors expanded at a very fast rate throughout the post-war period. By 1959, however, the greater convenience of oil for general industrial use and its growing availability had been widely recognised and this trend was accelerated by the growing use of oils as feedstocks for Europe's petrochemical industry. Finally, oil products were selling at low market prices that compared very favourably with coal prices which had risen continuously since the end of the war and throughout the consequent period of fuel shortage. The cheap prices at which the oil companies were able to sell their products were mainly due to the rapidly expanding sales and an exceptionally rapid increase in crude oil production. World output of crude oil rose from 783 million tons in 1955 to 1,011 million tons in 1959, 1,259 million tons in 1962 and 1,447 million tons in 1964—an increase of nearly 85% in ten years. This in turn led to keener competition between the major producers. Competition was sharpest in the market for heavy fuels and was centred round those areas where there were a number of refiners or marketeers operating on a selective basis and who could take advantage of spot cargoes of oil from various sources or the low tanker freights that followed the excessive ordering of new tanker tonnage after the Suez crisis.

The plentiful availability and cheapness of oil accentuated and hastened the fundamental changes which had begun soon after the end of the war, whereby Europe ceased to be primarily an importer of finished products and developed its own refinery industry. Between 1957 and 1962 refinery capacity in the O.E.C.D. European area increased from 150 million tons per year to 271 million tons per year. By 1964, this figure had risen to 355·8 million tons per year. This represented therefore an increase in refinery capacity over the eight-year period of over 125%.

Existing plans and projects are expected to result in further rapid and substantial increases in refinery capacity, resulting in a total capacity for the Western European area by the early 1970s of over 500 million tons.

This development of Western Europe into a major refining area was followed, logically and following the same pattern as had already occurred in the United States, by a shift in the siting of new refinery capacity from coastal areas to areas of consumption. Thus, by 1959, about half of the new refining capacity planned or under construction was located at inland sectors, in particular in France and Germany. These refineries were projected along the routes of oil pipelines, the development of which came to a head in the late 1950s. As stated in the report published by the O.E.C.D. on the oil industry in 1964 'Where markets are small in total demand they are more likely to be fed from base-load refineries, that is, from large refineries which supply products to a number of area markets. The siting of these base-load refineries is usually on the coast

Table 25. *European Refinery Capacity (crude distillation) (by countries 1957–64)*

	Million tons per year at end year							
Country	1957	1958	1959	1960	1961	1962	1963	1964
Austria	2·4	2·4	2·2	2·2	3·3	3·4	4·5	4·7
Belgium	7·6	7·6	7·9	8·6	8·9	13·4	13·9	15·7
Denmark	—	—	0·2	0·2	1·2	1·7	3·3	3·4
France	33·5	35·5	37·4	40·2	43·6	44·5	51·0	62·6
West Germany	16·6	27·0	30·0	40·5	42·5	46·8	62·1	73·0
Greece	—	—	1·8	1·8	1·8	1·8	1·8	1·8
Ireland	—	—	2·0	2·0	2·0	2·0	2·0	2·0
Italy	30·0	30·0	34·7	45·7	48·9	56·8	69·1	73·0
Netherlands	17·3	17·3	17·9	22·8	23·2	24·9	26·8	29·9
Norway	0·1	0·1	0·1	0·1	2·2	2·2	2·5	3·2
Portugal	1·2	1·2	1·3	1·3	1·3	1·7	1·6	1·6
Spain	3·9	4·0	11·1	11·1	11·1	11·1	11·1	11·1
Sweden	2·5	2·1	3·2	3·2	3·2	3·2	3·8	4·0
Turkey	0·3	0·3	0·3	0·4	1·6	4·8	4·8	4·8
United Kingdom	34·9	38·2	44·1	50·0	51·0	52·7	57·6	65·0
Total	149·8	166·1	194·2	230·1	245·8	271·0	315·9	355·8

and in a position which gives the maximum economy in distribution to the various markets they serve. The new market-based refineries can often be of relatively simple design. Market demand produced patterns may not vary widely, and they will not therefore require the greater and more costly adaptability to be found in the large balancing refinery. Where refineries are of simple design, however, the crude and feedstock intake must be carefully chosen to give a reasonably good match in refinery yields to the products requirement of the market, and to minimise the problems of disposals of surplus products.'

Already by 1959, therefore, oil consumption in Western Europe had become large enough to justify pipelines of larger diameters designed to permit the bulk transport of crude oil from ports on the Mediterranean and North Sea to certain inland refineries constructed in areas of high industrial development. The economics of pipelines are such that they become financially attractive only if a continuous off-take can be assured; once this point has been attained they are highly competitive with all overland forms of transportation. By the end of 1964 the European crude oil pipeline network was as follows:

	Transport capacity (million tons per year)	Length (in km.)	Diameter (inches)
(i) In existence			
Zisterborg–Lobau (pre-war)	2	49	42
Wilhelmshaven–Wesseling	9–22	385	24
Finnart-Grangemouth	3·1	91	12
Rotterdam–Venlo–Wesseling–Wesel	85–20	300	24
Wesel–Gelsenkirchen	—	47	16
Le Havre–Petit Couronne	3·0	77	14
Genoa–Rho	1·1–2·1	129	12
Savona–Trecate	1·0–1·3	152	6/8
Milford Haven–Llandarcy	6	97	16
Lavera–Strasbourg–Karlsruhe	13–30	780	34
Karlsruhe–Ingolstadt–Neustadt	8–15	270	26
Wesserling–Kelsterbach	2·0	150	24
Genoa–Aigle–Ingolstadt	16–18	900	12–16
Genoa–Cremona	—	150	12–14
Malaga–Puertollano	—	283	—
Kuibishev–Schwedt–Bratislava–Szaszhalombatta	—	4,500	24–39½
Donges–Rennes	4	100	12
(ii) Under construction or under study			
Adriatic–Southern Germany	—	—	—
Trieste–Schwechar	—	—	—
Genoa–Pavia	—	—	—
Thames–Merseyside	—	—	—
Marseilles–Dijon	—	—	—
Extension of the Russian Pipeline to Latvia, Czechoslovakia and the borders of West Germany	—	—	—

Considerable interest was also being shown at this time in the possible development of product pipelines, of which there were only three in operation in Europe, that is, in France, Spain and the United Kingdom. The long-distance transport of heavy oils does however present a number of technical difficulties, and it is therefore probable that for the near future, at least, pipelines designed to transport the higher grades, that is, gasolines and middle distillates, will predominate. Further petroleum product pipelines are, however, projected in France, Germany, Spain and the United Kingdom. Thus, by the end of 1966 the total number of product pipelines had risen to nine in the Community countries above. With the rapidly rising demand for oil all over Europe, further pipeline construction can be expected to feature prominently in the plans of both the oil companies and national Governments.

A survey of the pattern of consumption of fuel oil and gas diesel oil by country in the O.E.C.D. area in 1964 showed that out of a total inland consumption of 197 million tons of gas oil and fuel oil just under 70 million tons were taken by general industry. Public electricity plants accounted for a further 25 million tons, the iron and steel industry for about 12 million tons and transport (mainly railways and coastal and inland navigation) for about 4 million tons. It is significant that in the five year period from 1959 to 1964 consumption of gas oil and fuel oil in general industrial and electricity generating sectors increased by more than 100%.

While oil consumption and refinery capacity has increased dramatically in certain areas between 1957 and the present day, often to the disadvantage of other fuels, notably coal which was faced with grave social, economic and manpower difficulties, the overall energy situation in Eastern Europe maintained a satisfactory equilibrium due to the fact that, fundamentally, energy availability remained marginally below demand. While oil consumption in Western Europe (excluding the United Kingdom) rose between 1957 and 1964 from about 120 to over 330 million tons of coal equivalent, the corresponding figures for Eastern Europe (excluding the U.S.S.R.) were 14·5 and 33·0 million tons respectively. In the U.S.S.R. oil consumption rose by 68 million tons to 221 million tons. During the same period consumption of solid fuels in Western Europe fell from 380 to 326 million tons, while in Eastern Europe there was an increase from 225 to 298 million tons and in the U.S.S.R. from 352 to 454 million tons. (Tables showing the development of internal consumption of commercial sources of primary energy in Europe from 1950 to the present day have been included at the end of Section 2.) Any attempt at a comparison between the overall energy situation in Western and Eastern Europe however is inevitably vitiated by the totally different market forces controlling the two areas. Whereas in the West direct control or interference in the energy sector by Governments is normally limited to assuring adequacy of supplies, avoidance of social or market disturbances as a result of a temporary surplus of any one source of fuel and arranging long-term security of supply (i.e. new energy-producing industries such as atomic energy), in Eastern Europe production and imports of energy are rigidly controlled by the Governments so that availability is geared as closely as possible to actual demand with the result that prices are artificially pegged and are not subject to the same competitive market forces as elsewhere.

The change-over from the era of energy shortage to one of abundance of fuel supplies gave Europe a basic two-fuel economy. Between them, coal and oil in 1964 accounted for over 84% of Western Europe's energy supplies. But, while coal in 1964 still accounted for 43% of the total and thus remained, marginally, the largest single source of energy, it was evident that this place would shortly be usurped by the oil industry. Natural gas and hydro-electricity contributed in 1964 no more than 16% to Western Europe's total energy requirements, while atomic energy will not be in a position to make a substantial contribution to Europe's energy needs before the 1970s. For the immediate future therefore coal and oil will constitute Europe's main source of energy and it is to the effect of the upsurge in oil consumption upon the less adaptable European coal industry that we must now turn.

The re-organisation of the European coalmining industry

The slackening in the rate of industrial activity in Western Europe which began in 1957 and continued till 1959 together with a succession of mild winters and the growing popularity and availability of oil brought about a complete change in the position of the coalmining industry. From 1945 onwards the coal industries had been working under instructions to produce every available ton of coal in order to enable the rapidly mounting energy requirements of Western Europe to be met. As a result there was a steady increase in output which reached its highest level of 592 million tons in 1957 but had fallen by 1960 to a total of 451 million tons. Four years later, in 1964, output was down to 445 million tons, while in 1965 there was a further drop of 17 million tons. From 1958 onwards there was a marked change in emphasis from maximum production to a level of output which would facilitate the production of those types of coal most suited to general market requirements. No similar development occurred in Eastern Europe and the U.S.S.R. where, as a result of rigid market disciplines and a continuing basic shortage of fuel, coal production continued to increase, rising from 327 million tons in 1957 to 515 million tons in 1960, 568 million tons in 1964 and 587 million tons in 1965.

But the disequilibrium in the energy market of Western Europe and the rise in pithead stocks that resulted from it cannot be entirely attributed to the fall in demand for coal and its replacement by other fuels. The problem had been vastly aggravated by the rise in coal imports from outside Western Europe, notably the United States. This trend towards increased coal imports from outside the European area dated back to the spring of 1956 when the level of economic activity was particularly high and European coal production was insufficient to meet requirements. At that time consumers in many countries increased their import orders from the United States in order to cover their requirements for current consumption and to build up their stocks. At the end of 1956 the political uncertainties associated with the Suez crisis led to a further sharp increase in purchases of American coal. The dearth of shipping on Transatlantic traffic caused a spectacular rise in Transatlantic freight rates (*i.e.* from $8 per ton from Hampton Roads to Antwerp/Rotterdam in January 1956 to $17 per ton by December 1956) with the result that importers, anxious to ensure steady and regular supplies, decided in many cases to conclude long-term contracts for purchases of American coal and sought to make charter arrangements considered to be favourable at that time. Once the Suez crisis had passed there was a radical change in the freight market with spot freights falling to between $3 and $5 per ton. This in turn led to a further round of speculative buying of American coal with the result that total imports of American coal shot up from 24 million tons in 1955 to 38 million tons in 1956, 45 million tons in 1957 and 31 million tons in 1958. Imports into Western Europe from Poland and the U.S.S.R. also rose in this period to a total of some 10 million tons per year. Between 1960 and 1964 imports of coal from outside the European O.E.C.D. area averaged only just under 58 million tons a year and contributed dramatically to the continuing crisis in the European coal industry.

This massive coal import programme coupled with the slowing down in the rate of economic activity and increased competition from oil led to a sharp and serious increase in the level of undistributed stocks of coal:

Table 26. *Stocks of Hard Coal at Pithead in Western Europe (in million tons)*

Country	Dec. 1957	Dec. 1958	Dec. 1959	Dec. 1960	Dec. 1961	Dec. 1962	Dec. 1963	Dec. 1964
Belgium	1·4	6·9	7·5	6·6	4·4	1·4	0·5	1·5
France	4·6	7·4	11·0	13·2	11·5	8·6	6·1	5·7
W. Germany	0·9	9·5	11·7	7·1	8·3	6·1	3·8	8·7
Italy	—	—	0·1	0·1	0·0	0·0	0·1	0·1
Netherlands	0·4	0·8	0·9	0·7	0·6	0·5	0·4	1·0
United Kingdom	8·8	20·1	36·2	29·7	21·8	25·8	19·9	20·7
Total	16·1	44·8	67·4	57·4	46·6	42·4	30·8	37·7

Thus, by the end of 1959, pithead stocks of coal had reached the unprecedented total of nearly 70 million tons and this was parallelled by a substantial increase in consumer stocks.

Although the situation improved between 1961 and 1964 as a result of the cold winters of 1961/2 and 1962/3 and the general cutback in coal production, by 1966 the fall in demand for coal, particularly in Western European continental countries, had again begun to outstrip the reduction in output so that pithead stocks resumed their upward trend. By the middle of 1966 stocks in the Community countries alone had again passed the 30 million ton mark.

In Eastern Europe pithead stocks remained at a low level and any minor imbalances arising from unforeseen climatic variations were corrected by either increasing or running down consumers' stocks. Thus, in the U.S.S.R., pithead stocks at the end of 1959 were less than 11 million tons, that is, under 2% of annual output. In Poland the stock figure at the end of 1959 was just over 800,000 tons, or 0·8% of annual output, while elsewhere in Eastern Europe stocks seldom, if at all, exceeded 5% of output. Similarly, throughout the early 1960s, pithead stocks of coal in the Eastern European countries and the U.S.S.R. remained at extraordinarily low levels, amounting at the end of 1964 to less than 150,000 tons in Poland and some 200,000 tons in Hungary.

As a result of the serious imbalance and the depression facing the coal industries of Western Europe, Governments and producers were faced with an urgent need to find and implement corrective measures. In Belgium, which was the country most seriously affected by the coal crisis, the Government decided in February 1958 to re-introduce the licensing system for coal imports which had been abrogated only some six months earlier. At the same time the Belgian Government recommended to importers of American coal that they should spread out over a longer period the tonnages they had undertaken by contract and by chartering to take up in 1958. Two years later, in 1960, coal import quotas, for countries outside the European Community, as well as a tax of 5% on imported coals, were introduced. The Government also increased its financial assistance to the Belgian coalmining industry by both direct and indirect means. Direct financial assistance had been given to the industry since 1945 in order to assist it in its post-war reconstruction programme and reached a peak in 1953 when the total amount was some £17·5 million. This however subsequently fell rapidly, largely as a result of the provisions of the Treaty of Paris establishing the Coal and Steel Community which expressly forbad direct financial assistance by governments to their national coal and steel industries. Even so, however, by 1964 direct subsidies were again in the region of £7 million a year. During this period, the Belgian Government also gradually increased its contributions in the form of indirect aid (*i.e.* miners' pensions fund) to a total of some £40 million by 1964. Finally, the Belgian Government, in 1960, imposed a tax on fuel oil—introduced as a diminishing tax, starting at a rate of 8*s.* 7*d.* per ton—which was reviewed and fixed at 5*s.* per ton in 1962. The proceeds of this tax were used during the first two years to promote Belgian coal exports.

Government action was also taken in three other Community coal-producing countries, that is, France, Germany and the Netherlands. In Germany a law was passed in 1959 imposing a duty-free quota for non-Community countries; any imports in excess of this quota were made subject to a duty of £2 per ton. The German Government also guaranteed loans to the coal industry to secure the cancellation of contracts to import American coal. At the same time the German Government contributed towards increased wage costs by means of a payment made directly to the miner for each shift worked, amounting in total to about £14 million per year; this was followed, in 1962, by a further direct grant of about £22 million, the money being provided out of the proceeds of the tax on fuel oil. In addition, the Government also agreed, in 1960, to finance a rebate on coal transport charges at a cost of about £8·5 million per year. Even more sweeping measures to assist the German coalmining industry were introduced in September 1963 under a new law establishing a rationalisation cartel or agency. The object of the agency is to promote and assist the rationalisation of the German coalmining industry so that, in the words of the law, it may 'assume its position in the Federal German economy'. The agency arranges for colliery companies which close pits for rationalisation purposes to be paid about 45*s.* per ton of abandoned capacity, half of this sum coming from public funds. Furthermore, in conjunction with the rationalisation agency, the Government guaranteed a fund of about £133 million to finance the rationalisation of the industry and granted tax concessions to encourage rationalisation measures. Other measures envisaged in the law include restrictions on oil imports from Eastern European countries and the Soviet Union as well as the establishment of statutory minimum stocks of oil.

During 1965, as part of a programme allegedly designed to help the industry to maintain an annual level of sales of around 140 million tons, the Federal Government introduced measures to promote the use of coal by giving tax relief on the capital cost of the construction or extension of power stations committed to coal for 10 years and by grants towards coal-fired group heating installations. The Government also secured an undertaking from the oil companies to confine their expansion to the rate of growth in the overall demand for energy without encroaching further on the market for coal; placed the oil companies under a statutory obligation to notify details of existing refineries and pipelines and projects for new construction or extension; and passed legislation to introduce, starting in 1966, the building up of stocks of oil products by 1970 to a level representing 65 days' average imports of crude oil.

Towards the end of 1965, with coal sales continuing to decline and with mounting stocks, the Federal Government and the Land Government of North Rhine/Westphalia agreed to assist with the financing of a programme of immediate measures designed primarily to stabilise pit-head stocks by together paying some £19 million towards the transport and stocking costs involved in the transfer over four years of 4 million tons of pithead stocks to places near to main consuming centres, and to compensate the industry for 4 shifts to be paid but not worked during the last two months of the year in order to reduce production by 2 million tons.

More recently still, in the summer of 1966, the German Government passed a Law providing for financial assistance to power stations undertaking to burn coal for a minimum period of 10 years. This measure was expected to result in an increase in coal consumption at German power stations of some 2 million tons a year for each year between 1966 and 1971 and thereafter to maintain coal consumption at power stations at some 43 to 45 million tons a year up to 1981. The cost to the German exchequer was estimated at DM 1,500 million.

By the end of 1966, both Government and industry had abandoned any pretence of endeavouring to maintain coal production at a level of 140 million tons. Output, it was freely recognised, would fall from the 1965 level of 135 million tons to a figure of about 100 to 115 million tons by 1970, with the likelihood of still further falls thereafter.

In France and in the Netherlands coal imports from non-Community countries have always been subject to careful Government procedures. In France all imports have to be approved by a government agency known as ATIC (Association Technique de l'Importation Charbonnière) which also supervises the execution of the contract and has an exclusive right to arrange transport. The French coalmining industry has also for some years received direct financial support from its Government. In 1956 the Charbonnages de France was relieved of £7 million interest charges and in 1958 received State aid to the extent of £14 million as compensation for Government-imposed restraint on coal prices. In 1959, the Government converted £265 million of advances to the Charbonnages into long-term Government stock; the Charbonnages now pay a nominal 1% on this part of their borrowings and are thereby relieved of about £8 million a year in interest payments. In 1960, the Government made a grant of nearly £4 million as a contribution towards the re-organisation of the French coalmining industry. This was increased to £11 million in 1961, to £15 million in 1962 to nearly £50 million in 1963, £34 million in 1964 and some £44 million in 1965. At the end of 1964 the Charbonnages de France signed an agreement with the French electricity authority, under Government auspices, for the supply to the power stations at the official list prices less 2% quantity discount of an additional 1 million tons a year over the next five years.

Indirect financial assistance was also given to the coalmining industries of the countries of the European Coal and Steel Community by the High Authority. This has taken two main forms: firstly, by means of loans at favourable rates of interest for new projects submitted to, and agreed by, the High Authority. Secondly, by making grants towards re-adaptation in the coal industry, covering special redundancy pay, training mineworkers for new jobs and resettlement allowances. Grants for re-adaptation were given on condition that Governments of the countries concerned matched these by giving subsidies to the same amount. Up to the end of 1965, the High Authority had allocated some £17·5 million to the Community coal industries. Of this total about £2·5 million was spent on emergency measures to deal with the coal crisis, that is, the financing of pithead stocks and special allowances for Belgian miners put on short-time working. The High Authority also authorised the Belgian Government—authorisation being required under the Treaty of Paris—to give direct subsidies to its mining industry totalling £14 million—of which only £2 million was repayable—in three years up to 1961.

In the United Kingdom, Government assistance took the form of a ban on U.S. coal imports and supervision of the level of oil imports. Government support was also an important factor in maintaining the high and expanding level of coal consumption by the C.E.G.B. Above all, the 1965 Coal Industry Act provided for a capital reconstruction of the industry, which relieved it of £415 million of its capital debt. This reduced the interest charges in the Board's accounts for 1965–66 by £21·5 million and depreciation charges by £14·1 million. It is important to remember, however, that the coal industry's accumulated deficit had risen to a large extent as a result of the high investment expenditure incurred in developing pits and production capacity during the late 1940's and early to middle 1950s to meet the high levels of demand then envisaged in official circles as well as to the cost of selling imported American coal (during the middle 1950s) at the ruling lower prices of British coal.

These forms of governmental assistance were paralleled by energetic and wide-reaching efforts by the coalmining industries themselves to face up to competition from oil and to re-adjust themselves to the changed market conditions. In addition to wide-reaching measures to modernise the image of coal by means of publicity and sales promotion campaigns, solid progress was made in the vital field of increased productivity. This was achieved by closing uneconomic pits and the extension of mechanisation and automatic techniques. Pit closures constituted the negative side of the rationalisation measures taken by the coal producers. They were however inevitable in the conditions facing the industry from 1958 to 1961, when a number of pits in Western Europe were compelled to put miners on short-time working.

Table 27. *Coal not produced on account of short-time working in Europe (in thousand tons)*

Country	1958	1959	1960	1961	Total 1958–1961
Belgium	2,133	5,701	3,090	771	11,695
France	—	380	1,841	294	2,515
Germany	4,134	6,183	835	104	11,256
Total	6,267	12,264	5,766	1,169	25,466

With pithead stocks rapidly reaching the maximum supportable physical and financial limits in Western Europe and with the increase in short-time working in certain countries, measures were taken to close uneconomic pits despite the reluctance of the coal producers. The closure of a mine generally results in the permanent loss of its resources; furthermore, it takes some ten to fifteen years for investments in new mines to materialise fully and some twenty to fifty years for the investments to be amortised.

The progress achieved by the European coalmining industry in positive rationalisation, that is, increasing productivity, is as follows (in kilograms per manshift underground):

Year	Belgium	France	West Germany	Netherlands	United Kingdom	Czechoslovakia	Poland
1957	1,150	1,683	1,585	1,499	1,598	—	—
1959	1,265	1,716	1,845	1,619	1,639	1,869	1,736
1960	1,577	1,798	2,126	1,833	1,950	1,773	1,793
1962	1,818	1,922	2,459	2,117	2,293	1,642	1,931
1963	1,820	1,958	2,618	2,137	2,426	1,656	2,007
1964	1,753	2,046	2,604	2,208	2,439	1,649	2,080
1965	1,852	2,039	2,705	2,253	2,554	1,712	2,154

Thus, in the space of eight years, productivity increased by 61% in Germany, by 54% in the United Kingdom, by 52% in Belgium and by 44% in the Netherlands.

As a result of the new market pressures and the changed conditions governing the European energy market, hard coal production from 1957 onwards was gradually reduced in nearly all Western European countries. Production of brown coal however, which is mainly consumed by power-stations nearby or on the site of brown coal deposits, remained at about the same level and in some cases showed a slight increase. In

Table 28. *Production of hard coal in the main producing countries (in million tons)*

Year	Belgium	France	West Germany	United Kingdom	Poland	U.S.S.R.
1957	29·2	56·8	149·7	227·2	94·1	328·5
1959	22·8	57·6	141·8	209·5	99·1	365·4
1960	22·5	56·0	142·3	196·8	104·4	374·9
1962	21·2	52·4	141·1	200·6	109·6	386·4
1963	21·4	47·8	142·1	198·3	113·2	392·0
1964	21·3	53·0	142·2	196·7	117·4	409·0
1965	19·8	51·3	135·1	190·6	118·8	427·7

Table 29. *Production of brown coal in the main producing countries (in million tons)*

Year	West Germany	Yugoslavia	Czechoslovakia	East Germany	Hungary	U.S.S.R.
1957	96·8	16·7	51·0	212·6	18·9	135·0
1959	93·4	19·8	53·7	214·8	22·6	141·4
1960	96·2	21·4	58·4	225·4	23·7	138·3
1962	101·3	23·5	69·5	244·9	25·3	131·0
1963	106·7	26·1	73·3	254·4	26·8	140·0
1964	110·9	28·2	74·5	256·9	27·4	145·0
1965	101·9	28·8	72·3	251·1	27·1	150·3

Eastern Europe production of both hard and brown coal increased steadily throughout the period.

By 1960 the long-term plans for hard coal output in Eastern Europe were, in their turn, beginning to show an increased emphasis on the output of coking and gas coals and on measures designed to increase productivity and reduce costs. Already measures were being taken to review plans for the development of lower-grade brown coal, of which, as Table 29 shows clearly, Eastern European countries are by far the more important producers—and where, unlike most Western European countries, brown coal also constitutes an important domestic fuel. This seemed to suggest that while the policy of maximising output had not yet been abandoned—as it had been irrevocably abandoned in the West—it was nonetheless being modified, especially in the U.S.S.R., to take greater account of qualitative and cost considerations and the greater availability of lower cost energy.

Coal's declining share of Europe's energy market

It is a measure of the success of Western Europe's coalmining industries in meeting the challenge from oil that between 1961 and 1963 absolute disposals and consumption of coal actually increased. At the same time the great upsurge in demand for energy has led to a sharp increase in oil consumption and oil's relative share of the energy market. In 1962 alone there was an increase in oil consumption in Europe as a whole of 52 million tons of coal equivalent—an increase of 18% over 1961. This increase was greater than that experienced in most preceding years; oil consumption rose by about 11% per year between 1950 and 1957 and by between 11 and 17% per year between 1958 and 1961. Between 1962 and 1964 the rate of increase accelerated to nearly 20%. Comparing the proportionate share of the market for each fuel in 1955—the last full year before the Suez crisis—and 1964, it is at once evident how the energy spectrum of Europe as a whole—but Western Europe in particular—is changing.

	Solid Fuels (%)		Liquid Fuels (%)		Natural Gas (%)		Hydro-Electricity (%)	
	1955	1964	1955	1964	1955	1964	1955	1964
Western Europe	75	50	16	38	1	2	8	10
Eastern Europe	90	83	5	9	4	7	1	1
Total	78	58	14	31	2	3	6	8

The increasing dependence of Western Europe on imported oil supplies and her metamorphosis from energy producer into energy consumer raised two major questions: firstly, whether Europe would at all times be able to rely on adequate and cheap supplies of oil and, secondly, whether Europe would be able to pay for these rapidly-growing energy imports. The questions were very much in the mind of the Hartley Commission, who had concluded in their 1956 report that 'the rapid increase in the dependence on imported energy with its inherent risks and the certainty of increasing prices both point to the urgent need for member countries to develop further their indigenous production of all forms of energy, having in mind both economic and security considerations'. Similar arguments were advanced by the Western European Coal Producers when they called for government assistance to help the coal industries in the struggle against oil often sold at barely economic prices as a result of the temporary surplus of supplies.[1]

The Coal Producers in their memoranda further pointed, with justification, to the untoward effect that uncontrolled competition from excess oil supplies would have on Europe's coalmining industries which by their nature were not geared to adapt themselves to the prices charged by the oil companies at the present time, but which might easily rise as steeply as they had fallen under the gradual pressure of increasing world demand for energy and rising investment, prospecting and royalty expenditure by the oil companies. To this the oil companies have usually replied that as for a number of technical and economic reasons Europe must inevitably come to draw increasingly on fuel supplies from overseas—if only because the total quantities of fuel required far exceed the production capacity of Europe's indigenous resources—the most effective way of achieving greater security lies in the diversification of sources from which energy supplies are drawn. The solution to this problem is one that confronts the Governments of all coal-producing countries in Western Europe today. The present indications are that in nearly all cases, that is, Belgium, France, Western Germany, the Netherlands and the United Kingdom, the Governments concerned, both for social, financial and security reasons, are not prepared to see any excessive reduction in coal production; as, however, demand for energy in all these countries is rising rapidly it follows that coal will become relatively less important. Thus, the German Government has recently taken measures designed to ensure

[1] See *Meeting Europe's Energy Requirements* (February 1963) and *An Energy Policy for Western Europe* (March 1966): memoranda by the West European Coal Producers and the National Coal Board.

the maintenance of an annual production at least in the medium term of the order of 100–115 million tons. In France, the recent Fifth National Plan provides for a reduction in coal output of some 10% or 5 million tons over the next five years. In Belgium and the Netherlands a significant reduction in coal production over the next decade is inevitable but this is unlikely to bring output below 12 million tons in Belgium and 7 million tons in the Netherlands. In the United Kingdom the Government White Paper of 1965 spoke in terms of a level of output by 1970 of between 170 and 180 million tons (although a review of the forward energy situation is now in progress). This question of determining the level at which indigenous coal production should be maintained is of course to a large extent the problem which is exercising the Governments of the member states of the Community in their efforts to arrive at a co-ordinated energy policy for their countries.

The move towards co-ordination of energy policies in the Community

The common market for coal and steel was introduced at a time of a more or less balanced economic situation and thereafter, until towards the end of 1957, the Community enjoyed a long period of boom conditions. At a time of rapid expansion and high demand, and when there was little to fear from external competition, the Community's industries were in a strong position, and the conditions with which the Community had to deal lent themselves without insuperable difficulty to the introduction and application of the new mechanisms and rules. These circumstances, and the political will of the member states in these early years to make the Community work, were favourable to the establishment of the common market for coal and steel, which was able to be brought into force relatively smoothly and even without recourse to some of the transitional safeguard measures provided for to mitigate the immediate impact of the liberalisation of trade. By 1956, therefore, the Community could justifibly claim that it had fulfilled the fundamental objective of its Governments by showing that European economic integration on a new basis of common interest and common institutions was a practical proposition, thus opening the way politically for the negotiations which resulted later that year in the signature of the Treaty of Rome and the establishment of the general common market and Euratom. By the late 1950s, however, there had come about a combination of a large number of factors which profoundly influenced the nature and operation of the Community. Prime among these was the resurgence of French nationalism, fanned by the accession to power of General de Gaulle, and manifesting itself by strong criticism of supranationality and the principle of the majority vote. In this policy the French had at that time the ready collaboration of the Germans who had always nursed a grievance against those provisions of the Treaty which were designed to control German predominance in the European coal and steel markets. The 1958/59 crisis of demand on the Community's coal industries, brought about, as we have already seen by the slackening in economic activity, the greatly increased availability of oil and massive coal import programmes, principally from the United States, coincided with these political developments.

The High Authority first tried to deal with the worsening coal market situation by the indirect methods of negotiations with the Governments, producers and other interested parties to secure agreed arrangements for remedial measures but found that its proposals were being whittled down, obstructed or flatly rejected by the Council of Ministers, consisting of Ministers of the six member states. Thus the Council failed to reach agreement on a Community system of financing stocks and on Community control of imports and flatly stated that there could be no question of limiting the sovereignty of member states in matters of commercial policy towards third countries. Faced with these circumstances the High Authority, early in 1959, concluded that the situation was so critical—pithead stocks of coal and short-time working were both still rising at this time—that the appropriate Community powers of intervention should be invoked to provide a solution. It accordingly submitted to the Council of Ministers proposals under Articles 58 and 74 of the Treaty of Paris for the declaration of a state of 'manifest crisis' and for the adoption of a Community plan to deal with it by the establishment under Community supervision of production quotas and import restrictions. After much discussion and negotiation with the Governments, during which the High Authority twice agreed to modify its plan to meet the objectives and stated requirements of individual member states but still could not gain acceptance of its proposals, the High Authority insisted on a vote, and the Council of Ministers formally rejected the proposals; France, Germany and Italy voted against them and the Benelux countries in favour.

This event marked a turning point in the Community's development. Although France and Germany were acting within their Treaty rights, their rejection of the High Authority's proposals was motivated mainly by political considerations and constituted in fact a refusal to accept the principles of supranational control written into the E.C.S.C. Treaty. In opposing the proposals the French Minister, supported by the German

Minister, bluntly stated that he could not approve a scheme which would give the High Authority new powers of control over the Community's coal industry; individual national Governments would be responsible for the political and social consequences of the High Authority's measures and they could not therefore allow control to pass out of their own hands. Thus, the economic problems of the coal crisis raised for the first time in an acute form the political issue between the concept of supranational or Community action and the interests of individual member states, and supranationality suffered a major defeat. The effect of this setback to supranationality was not to deprive the High Authority of its powers but to underline its dependence, in exercising them, on the co-operation of Member Governments. This dependence increased still more as energy policy and import policy, progress on which demands the unanimous consent of the Governments, have become increasingly important in Community affairs. The High Authority, while deploring the action of the Governments and receiving a shattering blow to its morale, was forced to recognise that aid to the coal industries to enable them to rationalise their production and generally adapt themselves to the changed market conditions was economically and socially essential. Since 1960 its efforts to find a solution to these problems have been pursued in the context of its work, in collaboration with the European and Euratom Commissions, on the formulation of proposals for a common energy policy.[1]

The first proposals put forward by the Inter-Executive Energy Committee (composed of Members of the High Authority and the European and Euratom Commissions) in the early part of 1960 were based on a

[1] With attention being paid increasingly to national interests and the urgent problems of the coal industries in the changed conditions of the energy market, the Community's coal market has come to present a very different picture in practice from the common market as originally conceived by the authors of the Treaty of Paris. Although the Treaty prohibits 'subsidies or state assistance, or special charges imposed by the states in any form whatsoever', the German, French and Belgian Governments have for some years been giving, through independent national measures, substantial direct and indirect financial assistance to their coal industries. The coal market is also influenced by other individual arrangements which are out of keeping with the conditions of free trade, equal access by consumers to supplies on equal terms, and undistorted competition which are prescribed in the Treaty. In short, the Governments of the member states of the Community, mindful of each other's interests, have sought, by means of internal arrangements between themselves and by the acceptance of private undertakings between producers, to regulate their own internal markets and so prevent any undue disturbances that might otherwise have arisen if the provisions of the Treaty had been rigidly applied or enforced. This situation was in fact tacitly recognised by the subsequent Energy Protocol of 1964.

guidance price. The guidance price, to be determined by the Council of Ministers, was to be the price to which the Community's coal industries should eventually adapt themselves in order to be competitive with imported coal and oil. Stability in the market was then to be maintained by Government measures, such as compensatory duties, in the event of coal and oil imports entering the Community at below the guidance price level—either as a result of fluctuations in freight rates or because of temporary reductions in f.o.b. prices. The Committee also envisaged a period of adjustment in order to avoid economic and social disturbances arising from too rapid a contraction in the size of the coal industry. During this period the guidance price was to be fixed at a higher level than the current import price and protection was to be afforded to the coal industry by such means as tariffs and quantitative restrictions on coal and oil imports. It was not necessarily intended that the adjustment period, or the guidance prices operative during this period, should have been the same for all Community countries. In the course of its studies the Inter-Executive Energy Committee took a number of alternative prices, c.i.f. Rotterdam, varying from $11.50 and $14 per ton of steam coal, $13.50 to $15.50 for coking coal and $13.50 to $19.50 per ton of fuel oil. A series of calculations was then made to ascertain the extent to which prices would have had to be reduced in the various coalfields of the Community to meet competition from imported fuels at the prices indicated. These calculations showed that if guidance prices of $11.50 and $13.50 had been taken for steam coal and coking coal respectively then prices for corresponding Belgian qualities, to take only one example, would have had to be reduced by $5.25 and about $2; while a guidance price of $13.50 for fuel oil would have required a decrease in the price of Belgian coals in direct competition with oil of about $8 per ton.[1] These proposals did not however prove acceptable to the Council of Ministers. The next stage in the High Authority's efforts to promote a co-ordinated energy policy for the Community was a memorandum on energy policy for the Community which was transmitted to the Council of Ministers in June 1962. This memorandum proposed that there should be a common market in energy to be attained in three phases, consisting of a preparatory period ending in January 1964, a two-part transitional period running up to December 1969 and, finally, a common market in energy commencing

[1] The carefully worded and widely known view of the Inter-Executive Energy Committee was that only about 125 million tons of the Community's coal production could in the long-term be regarded as competitive with imported fuels. See *Etude sur les perspectives energetiques a long-terme de la Communauté Européenne*. Luxembourg, 1964. Chapter 15, p. 156.

on 1 January 1970. The purpose of the preparatory period was to determine the detailed measures to be taken during the transitional period, during which it was envisaged that taxes on oil would be progressively reduced to an agreed minimum level and that all coal and oil duties should be gradually abolished (except for those on imports from Communist countries). A system of subsidies for the coal industry was to replace the protection afforded by the duties and quotas on coal and oil imports. Thus in the final period when the common energy policy would have become a reality, there was to be free entry for coal and oil imports from outside the Community, a zero duty on crude oil and subsidisation of the Community's coal industry in order to maintain coal production capacity in excess of what it would normally have been in a free market. The memorandum contained no recommendation as to the future level of Community coal production or the amounts of subsidy to be paid to the coal industry. But, after months of wearisome discussions and amendments this proposal, like its predecessor, the guidance price scheme, had to be abandoned as a result of the failure of the Ministers of the Member States to come to any agreement. Many more months of fruitless discussion followed before, in April 1964, the Council of Ministers was finally induced to accept a Protocol of Agreement on energy policy which committed them to determine rules for a comprehensive common market for energy in the context of the then projected merger by 1967 of the three European Communities and the new single Treaty that this would involve. The Protocol did, however, provide for urgent action on the problem of coal. The Governments accepted the principle of State assistance to the coal industries and the High Authority was directed to submit proposals, in the framework of the Treaty of Paris, for a Community system governing State aids. The High Authority accordingly prepared these proposals in the form of a draft Decision under Article 95 of the Treaty of Paris. Broadly these proposals, which were to apply until the end of 1967, provided for advance notification by Governments to the High Authority of all proposed financial aid measures, and intervention by the High Authority if it considered that these aids were likely to cause distortion in conditions of competition between the coal industries; and authorisation by the High Authority, on application by Governments, of State aids towards rationalisation and related measures or, in exceptional cases, and subject also to the unanimous agreement of the Council, other forms of financial aid.

The next step along the long painful energy road was an attempt by the High Authority, early in 1966, to draw up a policy for coal, including a coal production objective. In so doing, the High Authority explained the limited scope provided by the Treaty of Paris for action by the High Authority itself and emphasised the extent to which possible solutions to the coal problem depended on the policies and the agreement of the Governments, whether for increasing aids to the industry, or for restricting imports from third countries.

At a meeting of the E.C.S.C. Council of Ministers in March of that year, the President of the High Authority, Signor Del Bo, stated unequivocably that, if the current situation for coal was bad, the medium-term prospects were no brighter. According to the High Authority's estimates, demand for coal in the Community by 1970 would be no more than 170 million tons, while production plans were expected to result in an output of up to 200 million tons. The choice that had to be made concerned on the one hand the rate of cutback in output which was immediately possible in the light of social and regional problems, and on the other hand the part to be played by coal in the years ahead. Future developments would be influenced particularly by the policy on imports and the level of State subsidies for the coal industry. The outcome of the meeting was that the Council recognised the serious situation of the Community coal industry and the urgent need to apply common solutions. On a proposal by the High Authority it decided to set up an *ad hoc* Coal Problems Committee of high-level government officials under High Authority chairmanship, which was to conduct exhaustive enquiries into recent changes in the Community coal market; to analyse the way in which the market was likely to develop, at least up to 1970, if no new measures were taken; and, within the objectives and provisions of the Treaty of Paris and the Energy Protocol of April 1964, to propose ways of resolving current and foreseeable difficulties on the Community's coal market and in particular the problem of coal surplus.

Shortly after the Council meeting the High Authority issued a 'Memorandum for the E.C.S.C. Consultative Committee on the Coal Objective for 1970 and Coal Policy', which was also submitted as a working paper to the *ad hoc* Coal Problems Committee. In this memorandum the High Authority put forward its view that for social and security of supply reasons the coal objective for 1970 should be 190 millions tons (compared with estimated sales possibilities in 1970 of 170 million tons if no new measures were taken), and that measures should be taken, on the basis of a common policy, to limit contraction to this level. To this end, national programmes should be co-ordinated in an objective accepted by the Six, who should work together to achieve it. On security of supply the memorandum argued that the 190 million tons was necessary in order that the proportion of total Community energy supplies in 1970 that would be provided from Community sources should not be too far removed from 50% of total supplies, and bearing in mind the need to preserve the possibility of setting a production objective for 1975–80 that would still make an effective contribution to overall security of supply. The memorandum then suggested lines of approach to the problem of securing (but without any sales guarantee to the producers) sales outlets for the 190 million tons. These included possible joint action for a more systematic use of quantitative restrictions on imports varying according to categories of coal (requirements of imported anthracite and steam coal should be reviewed, and imports of coking coal might be tied to obligations to use Community coal or to pay towards the maintenance of a security minimum of Community coking coal production); a system of specific aid on a Community basis to maintain the market for coking coal for the iron and steel industry; and ensuring for Community coal a substantial share in the power station market by compensating the cost advantage of competing fuels through price subsidisation, fiscal concessions or selective electricity tariffs.

The Coal Problems Committee submitted an interim report to the Council of Ministers for its meeting in May 1966. The report concluded that in the absence of new measures and in spite of an assumed reduction in production of 30 million tons there was likely to be in 1970 a surplus of coal supply over demand of the order of 1–7 million tons, and that 'in unfavourable conditions' the surplus could even reach 20 million tons. As to measures to be taken, there was no substantive agreement on restrictions on third country imports or on a Community financial mechanism to promote sales of Community coal (both of which approaches had been strongly urged by the German delegation, flatly opposed by the Italians and treated with great caution by the Dutch). Production objectives were regarded as dependent essentially on the possibilities for industrial redevelopment in advance of closures, which was the responsibility of the Governments; it was not possible therefore to fix a Community objective, although national objectives might be co-ordinated. The role of coal in security of supply could only be determined in the framework of overall supply of energy products. The only approach to facilitating the disposal of Community coal that appeared to be generally favoured at this stage was a strengthening of the measures taken or envisaged in the individual countries on the basis that these did not result in an increase in the price of energy or disturb the proper functioning of the common market. No agreement was reached on a possible solution to the problem, of particular interest to the Germans and the High Authority, of the alleged distortion of competition between the Community's iron and steel industries because of the substantial quantities of cheaper imported coking coal being used in certain countries, in particular Italy and the Netherlands. Thus the report ended by recommending to the Council to institute in the framework of the three Communities an overall study of security of supply for all energy products; to instruct the Committee to press on with work already begun (pursuant to the Energy Protocol of April 1964) on the problem of long-term coking coal supplies; and also to instruct the Committee to study production objectives in relation to sales possibilities; the scope of co-ordinating production objectives; and ways, in particular on a Community basis, of facilitating intra-Community trade in coal.

In discussion by the Council of Ministers on the Coal Problems Committee's report in May 1966 the main feature was a strong intervention by the German Minister deploring the lack of agreement in the Committee, calling for Community solutions to prevent the breaking up of the common market for coal and to ensure the disposal within the common market of all Community coal production; arguing that Community solidarity demanded support for the coal-producing countries by the others in return for assured supplies; and calling for urgent consideration of the problem of discrimination in coking coal supplies. The other Ministers confined themselves mainly to describing their national reorganisation programmes for the coal industry, and in the case of the Italian Minister, to casting doubt on the thesis that the use of third country coking coal by some steel industries and Community coal by others would lead to distortion of competition. Finally, the Council accepted the Committee's recommendation for further studies with particular reference to arrangements for the organisation of coking coal supplies, having regard to the question of promoting *intra*-Community trade in coal.

The High Authority accordingly proceeded to concentrate its attention on the problem of organising the market for Community coking coal in such a way as at least to maintain disposals (including *intra*-Community trade) up to 1970, and at the same time to remove, without harming the competitive position of the Community's steel industries in the world market, distortion in competition between those industries arising from different government policies on increasing supplies of low cost American coal which have resulted in a wide disparity in prices of coking coal for steel making in the individual countries. In July 1966 the Council of

Ministers considered proposals prepared by the High Authority and brought before it (with a number of reserves) by the Coal Problems Committee, for a Community system for subsidising (through payments to the producers) prices of Community coking coal (or coke made from it) to the steel industries to an extent necessary to approximate these at the point of consumption to those of third country products. The system provided for sliding rates of subsidy varying according to the distance between producer and consumer. The subsidies were to be paid to the producers by their Governments, but in the case of deliveries from one Community country to another (*i.e.* coal or coke involved in *intra*-Community trade) the subsidies would have been refunded to the Government concerned from a Community fund to be fed by contributions from all six member Governments according to a key to be determined. It was argued that the Community fund was justified on the basis of Community solidarity as expressed in *intra*-Community trade and in the provisions in the Treaty enabling all member States to benefit in times of scarcity from an allocation of Community coal production; this solidarity called for a sharing by all six Governments of the charges falling on the coal producing countries through maintaining supplies available for the steel industries of other member States.

The Council of Ministers failed however to reach agreement on these proposals; in spite of great pressure by the German Minister for a Community solution (the German steel industry had long maintained that it is suffering discrimination through being confined to Community coking coal while the Italian and Dutch steel industries benefit from American coal, and Germany supplies most of the coking coal and coke involved in *intra*-Community trade), the Italian and Dutch Ministers showed reserved attitudes and the French Minister, while ready to accept the subsidisation of coking coal prices nationally (with some Community co-ordination), flatly opposed a Community system on the lines proposed as a premature and partial measure and maintained that a solution should be sought in the context of the formulation of a common energy policy following the fusion of the Executives.

This outcome caused dismay in Community circles, especially as the German Government, while still not giving up hope of a Community solution, had been considering whether to introduce arrangements on a national basis for subsidising the prices of coking coal and coke for its own steel industry only; furthermore, an agreement reached recently between the French Government and the French steel industry in connection with a plan for the development of the latter provides for the subsidisation of coking coal and coke supplied to the French steel industry by the Charbonnages de France. The High Authority expressed great concern about the disruption of the Common Market that was threatened by this situation and tried desperately to promote a Community solution to the problem.

The object of the High Authority's proposals was to bring subsidies under Community supervision and to secure at any rate some degree of convergence of national measures in order to avoid disparities of such a nature as to disturb the functioning of the common market. The High Authority stated that these proposals were designed to achieve a proper balance between the demands of rationalisation and those of competition. They still, however, left a number of questions unanswered: for example, they left virtually untouched the mass of existing state aids, and they provided no frame of reference to indicate what would be likely to constitute unacceptable disparities. Even so, the Council of Ministers failed to reach agreement.

Following a round of bilateral talks by High Authority members with individual member Governments in October, proposals for a Community system were again placed before the Council at its meeting on 22 November; at the same meeting the Council was considering a whole set of measures and lines of action suggested by the High Authority for dealing with the problems of the Community steel markets. The proposals were still based on flat rate subsidisation of Community coking coal prices to approximate them to delivered prices of American coal, but as regards coal traded in between Community countries the financial arrangements would now be called 'a system of multilateral financial compensation between the six member states'. It was also proposed that the system should be subject to a limitation (not specified) on its period of application and to a limitation of the tonnages eligible for financial compensation. An example was given of such an arrangement showing tonnages based on 1965 deliveries and the cost of subsidisation. For the multilateral compensation covering *intra*-Community trade the example proposed that *one-half* of the cost of subsidisation incurred by a supplying country should be borne by the other five Governments according to a key to be determined; no indication was given of the amounts to be contributed by each. The Council did not reach a decision on the proposals, but instructed the *ad hoc* Coal Problems Committee to study the matter further and to submit for the Council's next meeting precise proposals for Community criteria to govern such subsidies and for a possible system of multilateral compensation covering *intra*-Community trade.

The French Minister was at pains to state after the meeting that his Government had not yet agreed to the principle of a Community mechanism, but in view of the modifications made to the proposals which it had rejected in July, it was prepared to accept the further study now to be undertaken and would decide its attitude when it had assessed the economic and financial implications of the proposals to be prepared by the Coal Problems Committee.

It seems probable that the conflict of interests between the coal-producing and non-coal-producing member States of the Community—which precluded any agreement with regard to the earlier proposals put forward by the Inter-Executive Energy Committee—will continue to bedevil and possibly obstruct any real progress. There is a considerable body of opinion within the Community which holds the view that the fusion of the three Executives—originally planned to take place early in 1965 but now unlikely to happen before 1968—may result in a more unified and incisive approach to the problem of achieving a measure of Community co-ordination of arrangements affecting coal and other forms of energy. The same observers believe that, as preparations are made for the merger of the three Communities—following upon the fusion of the Executives—a comprehensive common energy policy will be worked out on the basis of establishing, for inclusion in the new single Treaty, common rules governing competition between the different forms of energy, the granting of state aids, and commercial policy towards third countries. Be that as it may, it can be assumed that, despite the numerous difficulties arising out of the conflicting national interests of the six member states, it is only a question of time before a common energy policy is elaborated. It is inconceivable that, where the Community has found agreement and succeeded in reconciling conflicting interests in fields as complex and delicate as agriculture, they should fail to do so in energy policy. The lines of such an agreed policy, though still blurred in outline, are in any case beginning to take shape and will be based upon a policy of free imports for all energy requirements over and above the tonnage that can be reasonably economically produced by the Community's coal industries. This tonnage figure will certainly be substantially below the present level of production, and may well mean a further fall of 60 to 70 million tons by 1975. The disposal of this level of output would then be secured by subsidies on a national, or conceivably—although now rather unlikely—a Community, basis. In the latter case contributions by member States would be in proportion to their own national levels of production; some financial contribution might be made by the non-producing countries although the extent of such additional contributions must remain a matter of speculation. It is perhaps worth noting, before leaving this particular subject, that however protectionist the Community may be in other spheres, it is now well set on a liberal path in its proposed energy policies; the key to this dichotomy, no doubt, lies in the fact that one of the principal importers of energy in the Community is France, which stands much to gain and little to lose from adopting a liberal attitude in the strictly limited sphere of energy.

The Robinson report

Four years after the Hartley Report, a second detailed study was prepared by the O.E.E.C. Energy Advisory Committee, under the chairmanship of Professor Austin Robinson of Cambridge University. This Report,[1] which was published in January 1960 and which came to be known as the Robinson Report, reassessed the prospective energy requirements and supplies of Western Europe in the light of the developments that had taken place in the energy market since the Hartley Report was written.

The fact that significant changes had taken place was inescapable. In 1956, when the Hartley Commission had published its report, coal production in Western Europe was insufficient to meet the demands placed upon it. By 1959, although coal's share of the energy market of all countries belonging to the O.E.E.C. was still 52%—compared with 64% only two years before—the changing pattern of fuel consumption and the plentiful availability of oil had become apparent. The Robinson Commission began by reviewing the long-term forecasts with respect to energy requirements in 1965 and 1975 made by the Hartley Commission. Their conclusions were remarkably similar:

Table 30. *Comparison of energy consumption estimates prepared by the Hartley and Robinson Commissions* (*in million tons of coal equivalent*)

	1955	1965	1975
Robinson Commission Estimates			
Upper		1,050	1,425
Mean	777	1,010	1,325
Lower		970	1,225
Hartley Commission Estimates			
Upper		1,060	1,360
Mean	777	1,010	1,260
Lower		960	1,160

[1] *Towards a New Energy Pattern in Europe*, published by O.E.E.C. Paris, January 1960.

It will be seen that the energy consumption estimates for 1965 are almost identical, while the estimates for 1975 show an estimated increase of some 65 million tons in the case of the Robinson Commission. The fact that no increase was expected over the forecast made by the Hartley Commission for 1965 can be explained by the fact that there had been a slackening in the rate of general economic activity during the greater part of the period that elapsed between the preparation of the two reports and which, in the light of subsequent events, appears to have been accorded too much weight by the Robinson Commission. Turning to the trend in demand for individual forms of energy, the Commission stated that 'there are certain demands for energy that are rather rigidly specific to certain determined primary sources of energy. Thus demand for coal for carbonisation or for oil for use on internal combustion engines for transport purposes is rather rigidly specific. The choice between coal, lignite, natural gas, oil or nuclear fuels for the generation of electricity is by no means rigid and consumption will be attracted to one fuel or another in accordance with variations of the relative prices and the relative convenience and security of using them'. Expressed in quantitative terms, the Commission found that specific coal consumption was likely to rise only slowly from 170 to 210 million tons in 1955 to 200 to 225 million tons in 1965 and 225 to 250 million tons in 1975, while specific consumption of oil was expected to rise much more quickly from 50 to 70 million tons of coal equivalent in 1955 to 100 to 125 million tons in 1965 and 150 to 175 million tons by 1975. Consumption in sectors where demand was more flexible was expected to show increases of a similar order.

Table 31. *Estimates of specific and flexible demands for primary forms of energy in the O.E.E.C. area (in million tons of coal equivalent)*

	1955	1965	1975
More specific demands			
Coal, chiefly for carbonisation	170–200	200–225	200–250
Oil, chiefly for transport	50–70	100–125	150–175
More flexible demands			
For electricity generation	185	300–325	450–475
Other	322–372	295–450	325–625
Total	777	970–1,050	1,225–1,425

In this connection it is interesting to note that the Robinson Commission estimated that the total consumption of electricity in the O.E.E.C. area would rise from 358 TWh in 1955 to 700 TWh in 1965 and 1,200 TWh in 1975. This compared with low estimates of 600 TWh on 1,050 TWh and high estimates of 700 TWh and 1,400 TWh for 1965 and 1975 respectively by the Hartley Commission.

The Commission then proceeded to consider the contribution that could economically be made towards meeting these requirements from indigenous sources of energy. For coal, the Commission again put forward upper and lower estimates; the former based on the assumption (described as being unlikely in the Commission's opinion) that the availability of other sources of energy would rise less rapidly than had been anticipated; the latter based on the premiss that coal production could not be reduced beyond a certain level over a given period of time without running the risk of severe social disturbances. The Commission's estimates of coal production for 1965 and 1975 were as follows:

Table 32. *Estimates of possible hard coal production in the O.E.E.C. area (in million metric tons)*

		1965		1975	
	1955	Lower	Upper	Lower	Upper
European Coal and Steel Community	247	230	250	220	260
United Kingdom	225	205	220	200	225
Others	5	5	10	10	10
Total	477	440	480	430	495

It will be recalled that the Hartley Commission had estimated a hard coal production of 500 million tons by 1960, rising to 520 million tons by 1965 and remaining at that level till 1975. The probable contribution from other indigenous sources of energy by 1965 and 1975 was estimated by the Robinson Commission at 195 million and 300 to 310 million tons respectively.

Table 33. *Estimates of contributions by other sources of energy in the O.E.E.C. area (in million tons of coal equivalent)*

	1955	1965	1975
Lignite	30	45	60
Crude Oil	13	30	50
Natural Gas	7	25	50–60
Hydro-electricity	56	95	140
Total	106	195	300–310

Total supplies of energy from indigenous resources were therefore expected to reach between 670 and 715 million tons in 1965 and 780 to 915 million tons by 1975. This left a gap of some 275 to 375 million tons in 1965 and 410 to 635 million tons by 1975 to be filled by imports.

Table 34. *Estimates of total demand and of potential supplies of primary energy in the O.E.E.C. area (in million tons of coal equivalent)*

	1955	1965	1975
Estimates of total demand for primary energy	777	970–1,050	1,225–1,425
Mean of estimates	777	1,010	1,325
Potential indigenous supplies			
Coal	477	440–480	430–495
Lignite	30	45	60
Crude oil	13	30	50
Natural gas	7	25	50–60
Hydro-power	56	95	140
Nuclear energy	—	15–20	30–90
Other forms of energy	20	20	20
Total	603	670–715	780–915
Mean of estimates	603	690	850
Potential imported supplies			
Oil	146	260–310	380–500
Natural gas	—	5	10–75
Coal	28	10–60	10–60
Total	174	275–375	400–635
Mean of estimates	174	325	515
Total potential supplies	777	945–1,090	1,180–1,550
Mean of estimates	777	1,015	1,365

The Commission's estimates showed therefore an anticipated rise in oil imports from 146 million tons in 1955 to between 260 and 310 million tons by 1965 and between 380 and 500 million tons by 1975. No difficulty was foreseen in obtaining import requirements of this order of magnitude. With regard to the cost of such massive import programmes the report merely stated that 'in regard to the payment for the energy imports that we have indicated there are two factors that must be borne in mind. First, a significant part of the increase may come from the Sahara, which is within the French currency area and with which a balancing of imports by corresponding exports may be more than normally easy. Second, even where oil imports come from other sources, the net overseas payment, after allowing for interest and dividends on European holdings in the oil companies concerned, is considerably less than the import costs that appear in trade returns'. The relatively nonchalant dismissal of the cost of importing 510 million tons of oil was indeed a far cry from the very serious concern expressed by the Hartley Commission only four years earlier and can be attributed entirely to the radical changes that had taken place in the overall energy market conditions as well as in Western Europe's balance of payments position during that short period of time.

The Commission also examined the general problems of energy policy. It was not the task of the Robinson Commission to make proposals for an energy policy applicable to all the member States of O.E.E.C.—nor, indeed, did it attempt to do so. The Commission had been requested to state its independent view with regard to the 'economic, financial and political problems linked with the future development in the field of energy' and came down firmly in favour of a policy designed to secure an adequate supply of energy at the lowest possible cost with freedom of choice to the consumer. To this end Governments were enjoined to avoid the use of taxation of the fuel industry in such a way as to inhibit the use of cheaper forms of energy where this could be done without endangering security of energy supplies. The Commission rejected the view that the most satisfactory way of ensuring security of supply lay in a greater degree of self-sufficiency and argued strongly in favour of wider diversification of sources of supply as the best means of achieving this objective. The Commission also called, however, for larger holdings of stocks of oil and other imported fuels to assist in overcoming any possible short-term interruptions in supplies. As far as the European coal industry was concerned, the Commission considered a reduction in the size of the industry to be both desirable and unavoidable and implied that production for the industry should be progressively reduced to a level of some 450 million tons as soon as practicable.[1]

The Commission concluded their report with some hard-hitting general statements about energy policy in the O.E.E.C. area. These included three main points: first, that the predominant consideration in the formulation of a long-term energy policy should be a plentiful supply of low-cost energy with freedom of choice to the consumer; second, that uneconomic indigenous sources of energy should not be given protection

[1] *Towards a New Energy Policy in Europe*, *op. cit.* p. 79: 'Where gradual and progressive concentration of production is practicable, we hope that it will be adopted. But we doubt the wisdom of greatly delaying the necessary process of adjustment. We think that in the long run a smaller but technically efficient and prosperous industry will give better opportunities, wages, working conditions and security to those employed in it'.

or artificial encouragement; third, that the future development of the coal industry depended upon its ability to produce coal at competitive prices and that this could only be ensured by a concentration of productivity (an overall reduction in output) where the prospects of increasing productivity were greatest.

It was in these polished and theoretically unobjectionable conclusions and the emphasis they laid upon the need for the coal industry to streamline its output in order to try and develop its competitive position *vis-à-vis* imported fuels, that the report by the Robinson Commission was in starkest contrast with recommendations made only four years earlier in the Hartley Report. The Robinson Commission was widely criticised in coal industry circles on the grounds that it had reached these conclusions at a time when the fuel surplus was still a comparatively recent phenomenon and before it had become evident that there had been a structural change in the European energy market. The Robinson Commission in fact brandished the torch of free trade in coal well before the Inter-Executive Energy Committee of the European Executives in Brussels and in Luxembourg. This excessive haste in writing down the European coalmining industry coupled with the somewhat doctrinaire readiness to accept a very great degree of dependence upon imports of energy from outside the O.E.E.C. area, constituted two conclusions, perhaps too hastily arrived at, in a report which, while arousing a considerable degree of controversy and even indignation in coal producer and trade union circles, had the merit of being the first major official publication to describe in such categorical terms the changing pattern of Europe's energy requirements.

The 1966 O.E.C.D. Energy Committee's Report

Six years after the Robinson Report, in 1965, the Energy Committee of the O.E.C.D. decided, with approval of the Council, to undertake a further general study of the energy situation in the whole of the O.E.C.D. area. This report, which was published in mid-summer 1966, was entitled simply '*Energy Policy—Problems and Objectives*'.[1] Unlike its predecessors (*i.e.* the report by M. Louis Armand and by the Hartley and Robinson Commissions) this latest report by the O.E.C.D. was the work not of independent experts but of the Energy Committee itself.

The Energy Committee estimated that primary requirements of energy in the O.E.C.D. European area would rise from 1,073 million tons of coal equivalent in 1964 to 1,370 million tons in 1970 and 2,050 million tons in 1980.

[1] Published by O.E.C.D., Paris, 1966.

These forecasts were based on forward sector estimates and on the assumption that the Gross National Product (G.N.P.) throughout the European O.E.C.D. area would grow at an average annual rate of 4·5%, compared with an actual growth rate of 5·2% between 1950 and 1964.

Estimates of Total Primary Energy Requirements 1970 and 1980 O.E.C.D. Europe

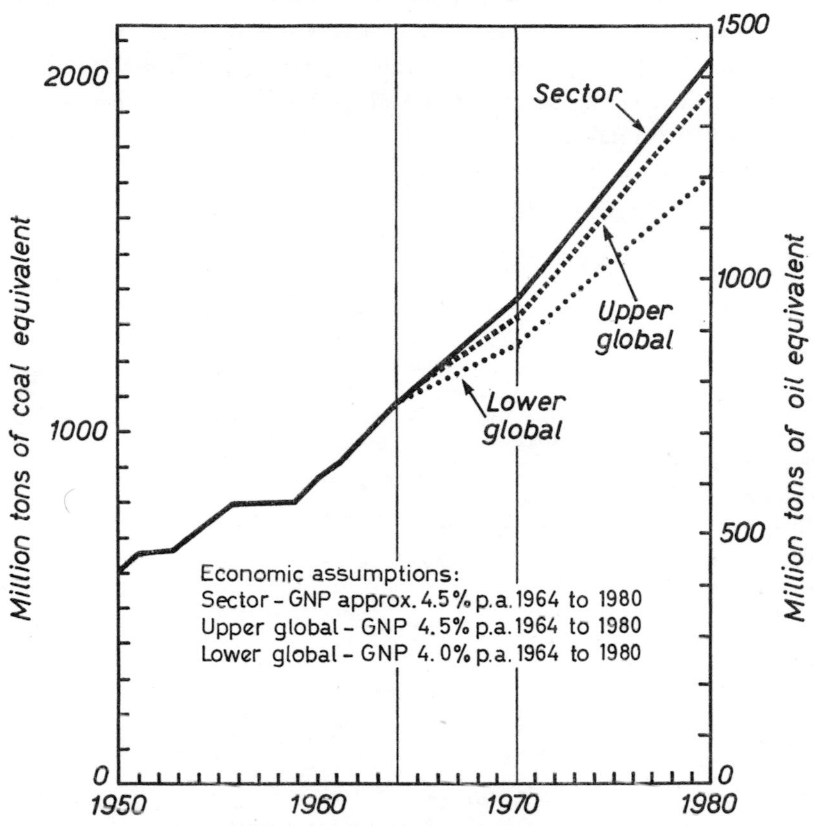

Source: O.E.C.D. Energy Committee Report, 1966, *op. cit.*

As the authors of the report themselves admitted, however, a rate of 4·5% per year might well turn out to be excessively optimistic in a period when the demands of reconstruction and recovery from the destruction of the war years will no longer apply.

In addition to the analysis by sector, two global estimates of primary energy requirements were also made, based on annual G.N.P. growth rates of 4·5 and 4% respectively. Since the ratio between the average rates of growth and G.N.P. in the areas averaged about 0·8 between

1950 and 1964, a similar figure was used in the first of the Committee's global calculations and a slightly higher one in the second. The results, together with those obtained from the sector-by-sector analysis, are shown in the graph on p. 91.

In examining the contribution that might be made by indigenous fuels, the Energy Committee assumed that the trend towards rationalisation of the coal industry and reduction of coal output would continue; that oil would continue to be freely available at about the present level of prices; and that consumption of natural gas, as a result of imports and further discoveries in the European areas, would grow steadily to reach some 10% of total energy consumption; and that in most countries, nuclear power would be competitive with conventional sources of energy by the end of the period under review.

Table 35. *O.E.C.D. Europe: Energy Supply and Demand expressed in terms of the main primary sources*

	Million tons of coal equivalent				
	1950	1960	1964	1970	1980
Total Requirements	600	867	1,073	1,370	2,050
Supplied from Indigenous Sources	513	582	584	565–630	725–950
Coal	461	473	447	375–410	300–335
Lignite	26	33	38	35–40	40–55
Oil	8	23	30	35–40	30–55
Natural Gas	1	16	23	50–80	100–160
Hydro Power	17	36	40	50	70
Nuclear Power	—	1	6	20	185–225
Net Imports	87	285	489	805–740	1,325–1,100
Oil	79	258	447	—*	—*
Solid Fuels	—	—	—	—*	—*
Natural Gas	8	27	42	—*	—*

* Partly a function of energy policy.

The most obvious and striking feature of the table is the extremely rapid rate at which the European O.E.C.D. area is becoming dependent for its energy supplies upon extra-European sources. Thus, whereas in 1950 energy imports were equivalent to less than one-seventh of total requirements, by 1960 they had risen to about one-third, and by 1964 to only a little under one-half; by 1970 they are expected to account for over 55% and in 1980, if we accept the higher of the O.E.C.D. estimates, over 64% of total requirements will have to be imported. The problems posed by such a degree of dependence on outside suppliers and, more particularly, on the politically unstable oil-producing countries of the Middle East, are evident. Equally, the effect of the mass of energy imports upon the balance of payments of many European countries is both onerous and apparent. These two considerations are not examined in any great depth of detail in the Energy Committee's report, which appears to rely upon extensive diversification of oil supplies as the best means of avoiding the dangers of excessive dependence upon one or two sources of supply. Similarly, the argument that indigenous coal production should be maintained at present or near-present levels, in order that it might constitute some security of supply, is dismissed as too costly and unproductive in operation (although the report does admit that some Governments in coal-producing countries have taken steps to maintain their coal production for financial and economic, as well as social, reasons). Indeed, it is quite the opposite philosophy that prevails throughout the report, which states in its conclusions:

'Far from revealing signs of imminent shortage of energy, our analysis of energy and demand suggests that ample supplies will be available for O.E.C.D. countries at reasonable costs to support continued economic growth up to 1980 and beyond'—'Establishing objectives of energy policy', the report continues 'working towards their achievement by employing suitable measures, and testing the adequacy and effectiveness of such measures should be a continuous process. A common energy policy cannot at present be the aim of the O.E.C.D.; the same basic necessities, however, underlie the energy policy issues facing most or all of the O.E.C.D. Member countries. It may be hoped that in framing their policies, governments will direct their efforts to measures which are most likely to lead to resources being used to the best advantage, to promote economic growth and to benefit rather than harm the interests of other countries. In this respect, O.E.C.D. can do useful work in continuing to provide the forum where Member countries can discuss all aspects of their energy policies, and work towards the achievements of their common objectives.'[1]

As in the case of the Robinson Report therefore, the 1966 report of the Energy Committee does not set out to formulate a common energy policy for the whole of the European O.E.C.D. area. The report is, however, intended to give some measures of guidance in the framing of energy policies and as such will no doubt be widely read and wield some

[1] 1966 O.E.C.D. Energy Committee report, *op. cit.* p. 143.

considerable influence. The report's apparent readiness to accept a rapid continuing decline of the coal industry in Western Europe and its curt dismissal of the need to maintain an indigenous coal industry, while in the direct tradition of the Robinson Report, make grim reading for the European coalmining industries. It also endorses the view that the majority of Western European Governments have accepted their countries' long-term dependence upon extra-European energy supplies.

The development of nuclear power

'Within a nuclear reactor the controlled fusion process yields neutrons and related forms of radiation, fission products and heat. Neutrons, over and above those needed to carry on the chain reaction, are chiefly of value for producing fissionable isotopes; and in so-called "production" reactors, as well as in breeders or convertors, they are used to produce plutonium from uranium 238 or uranium 233 from thorium. The fission products, sometimes unimaginatively referred to as radio-active waste, may be treated and particular isotopes separated for useful purposes. Heat, customarily thought of as the primary reactor product, may be used for generating electric power, for propulsion of naval vessels and, ultimately, space satellites; and for space heating and process heat at low or high temperatures'.[1] Despite the much-publicised maiden voyage of the Russian nuclear-powered ice-breaker, *Lenin*, or the American nuclear-powered merchant ship, *Savannah*, therefore, the vast bulk of experimental work for peaceful purposes up to the present time has been devoted to the application of nuclear energy for purposes of power-generation in power stations. By 1956, sixteen years after the invention by Professor Fermi of the heterogeneous reactor for purposes of uranium fission, three experimental nuclear power plants were in operation—in the U.S.S.R., the United States and the United Kingdom. The world's first nuclear power plant was put into service on 27 June 1954 in the U.S.S.R. near Moscow; it was equipped with a reactor operating on enriched uranium with a graphite moderator. The coolant was pressurised water, which passed through heat exchangers producing steam. The electricity was produced by a 5 MW turbo-alternator set. This was followed by the Calder Hall nuclear plant in the United Kingdom, which was first operated in October 1956 and the first experimental American power plant in December of that same year. By far the largest and most ambitious of these three, however, was the Calder Hall plant which was

[1] *Civilian Nuclear Power: Economic Issues and Policy Formation*, by P. Mullenbach. Published by the Twentieth Century Fund, 1963, p. 32.

intended to serve as a prototype for the subsequent construction of other nuclear power stations designed primarily for the purpose of producing electricity (the Calder Hall plant converted natural uranium into plutonium, which was required for military purposes. Nevertheless the heat released during this process could be, and was, used to generate electricity).

Despite the numerous forecasts made in the early 1950s about the growing demand for energy and Europe's absolute need for nuclear power in order to supplement its own inadequate energy resources, the history of nuclear energy throughout the 1950s and early 1960s was a relatively chequered one. This development, which was a disappointment to so many ill-conceived and exaggerated hopes, was due to two main and closely connected reasons: the unforeseen abundance of other fuels and their relative cheapness. To these may be added a third factor: the failure in the very early years to keep the cost per unit of electricity produced at a nuclear station at the level that had been forecast. In other words, not only was the cost per unit of electricity from conventional plants falling below, but that from nuclear plants was actually running above, the originally estimated gap. Already by 1955 at the First Geneva Conference, called to discuss, among other matters, the diminishing resources of conventional fuels, the conclusion was reached that there was no immediate urgency for nuclear power building programmes[1] and that the main problem was not so much one of shortage of supplies and resources of energy as one of wide geographical dispersal. By 1965, eleven years, therefore, after the first experimental plant near Moscow had come into operation, nuclear energy was still of very minor importance within the overall framework of Europe's energy supplies, accounting for 4% of total supplies of energy in the United Kingdom, less than 2% in the Common Market countries and even smaller fractions elsewhere in Europe. The Robinson Report, in a section devoted to nuclear energy,[2] had stated that with the capacity in operation or under construction in Western Europe at that time amounting to about 2·4 GW and a further 5·6 GW at the planning or negotiation of contracts stage, total available capacity by 1965 was unlikely to exceed 5·7 GW. The Commission was not able to give a firm estimate for 1975 in view of the widely differing views of its members, but suggested a range from 10 to 35 GW with a possible electricity production of 60 to 210 TWh. On the basis of the lower figure,

[1] *The World's Need for a New Source of Energy*. Proceedings, First Geneva Conference, Vol. I, p. 38.

[2] *Towards a New Energy Pattern in Europe:* Prospective developments in the field of nuclear energy, *op. cit.* pp. 49–56.

nuclear plants would represent less than 4% of the total electrical generating capacity added during the period.

In 1966, in their report *Energy Policy—Problems and Objectives*, the O.E.C.D. Energy Committee stated: 'The rate at which nuclear capacity is installed during the period under review will depend mainly on the growth of electricity demand and the trend of conventional compared with nuclear costs. Electricity demand is expected to continue to grow rapidly. Even allowing for the further decline in conventional generating costs, in some areas at least, nuclear power should be competitive for base load in most O.E.C.D. countries by about 1970. Costs, taken by themselves, would justify large programmes of nuclear installations in the 1970s, particularly in areas lacking access to cheap fossil fuel supplies. It is unlikely, however, that the rate of nuclear installation in the early years will be determined by straight cost comparisons alone. Utilities may be expected to proceed cautiously until adequate operating experience of nuclear stations has accumulated and cost estimates have been confirmed in practice.'[1] The report estimated that total nuclear capacity and generation during the period 1960 to 1980 would show the following rate of development:

	1960	1970	1975	1980
Nuclear output capacity (in GW)	0	10	40	90
Annual nuclear generation (in TWh)	0	70	280	630
Coal equivalent (in million tons)	0	47	171	436

The contribution made by nuclear power towards total generation of electricity in the European O.E.C.D. area was expected to rise from 0% in 1960 to 6·25% in 1970 and 29% in 1980.

In view of the different attitude and approach towards this new source of energy in various parts of Europe, we have thought it best to look at these developments by individual countries or, where more appropriate, by groups of countries.

The Common Market countries

In addition to their independent national nuclear programmes, the member States of the European Communities, in November 1956, that is, four months before the signature of the Treaty establishing the European Atomic Energy Community (Euratom) invited three experts, MM. Armand, Etzel and Giordani, to prepare a report 'on the amount of atomic energy which can be produced in the near future in the six countries,

[1] 1966 O.E.C.D. Energy Committee Report, *op. cit.* p. 58.

and the means to be employed for this purpose'. This report was published in May 1957 under the title *A Target for Euratom.*

The main conclusion of the report was that, confronted with Europe's rapidly growing energy import requirements, it was imperative to develop as rapidly as possible any indigenous source of supply that held out reasonable prospects of plentiful availability and ability to prove itself competitive: 'A second, still graver threat (the first was the balance of payments problem) is the evidence, provided by recent political events and the ensuing oil shortage, that even the availability of imported energy is uncertain. Oil already provides over a fifth of our countries' energy supplies. It is cheaper per calorie than imported coal, and it is more convenient to handle and use. It is therefore likely that most of the increase in demand which must be met by imports will take the form of oil. We cannot expect to obtain this oil from the Western Hemisphere because demand there is rising faster than production. The only region of the world capable of supplying these quantities is the Middle East, where a very high proportion of world oil reserves is located. The oil discoveries in the Sahara are promising, but they can hardly be expected to provide more than a fifth of our energy imports by the mid-1960s. Thus without nuclear power, Europe's dependence upon the Middle East is bound to increase. The Suez crisis has given us a warning of what this could mean. As the quantity of oil imported from the Middle East increases, there will be a corresponding increase in the political temptation to interfere with the flow of oil from that region. A future stoppage could be an economic calamity for Europe. Excessive dependence of our highly industrialised countries on an unstable region might even lead to serious political trouble throughout the world. It is essential that oil should be a commodity and not a political weapon. The European economy must be protected against the interruption of oil supplies, by finding alternative sources of energy to limit further rise in oil imports. Only nuclear power, providing Europe with a new source of energy, can achieve this'.[1] In view of this situation, the Committee recommended that the Community should develop a nuclear power station programme as a matter of urgency and seek to attain a target of 15 million kW of nuclear power capacity during the ensuing ten years. Less than a year later, however, the Euratom Commission itself cut back this figure to 4,000 MW capacity by 1965 and this was further reduced to 2,000 MW in 1960. Since then the Commission has formulated new proposals for a target of 40,000 MW by 1980.

[1] *A Target for Euratom*, pp. 17–18.

Details of all nuclear power plants in operation or under construction in the six countries by mid-1966 are given below:

Table 36. *List of nuclear reactors in operation or under construction in the Common Market countries by mid-1966*

Country	Site or designation	Type of Reactor	Type of Fuel	Capacity net MW	Date of entry into service
Belgium	Mol	PWR	UO2	10·7	1966/7
	Chooz (Franco–Belgian)	PWR	UO2	266	1966
France	Marcoule G1	GCR	U	100	1956
	G2	GCR	U	130	1959
	G3	GCR	U		
	EDF 1	GCR	U–Mo	62	1962/3
	2	GCR	U–Mo	213	1964
	3	GCR	U	476	1966
	EL 4	HWGCR	UO2	73	1967
	St Laurent 1	GCR	U-Mo	487	1967/8
	2	GCR	U–Mo	516	1970
	Bugey 1	GCR	U	488	1970
	Hospitalet (Franco–Spanish	GCR	—	500	1970/71
W. Germany	Karlsruhe	PWR	UO2	50	1965
	Juelich	HTGR	UO2	15	1966
	Gundremmingen	BWR	UO2	237	1966
	Kahl	BWR	UO2	25	1967
	Obrigheim	PWR	UO2	283	1968
	Lingen	BWR	UO2	240	1968
	Karlsruhe	LMCR	UO2	19·2	1970
	Niederaichbach	GCR	UO2	100	1970
Italy	Latina	GCR	U	200	1963
	Garigliano	BWR	UO2	160	1963
	Trino Verellese	PWR	UO2	255	1964
Netherlands	Dodewaard	BWR	UO2	47	1968

The key to the abbreviations for the various reactors (see also tables 37 and 38) is as follows:

PWR = Pressurised Water Reactor.
HBWR = Halden Boiling Water Reactor.
HWOCR = Heavy Water Organic Cooled Reactor.
HWR = Heavy Water Reactor
LMCR = Liquid Metal Cooled Reactor
HWGCR = Heavy Water Gas-Cooled Reactor.
OMR = Organic Moderator Reactor.
GCR = Gas-cooled Reactor.
HTGCR = High-Temperature Gas-Cooled Reactor.
SGHWR = Steam-Generating Heavy Water Reactor.
BWR = Boiling Water Reactor.
AGCR = Advance Gas-Cooled Reactor.

Despite the hopes and intentions of the fathers of the Euratom Treaty, which at the outset had the full support of the six national Governments and in particular of the French Government of the day and the 'European' element in French political life,[1] Euratom has not been able to co-ordinate the nuclear programmes, even for purely peaceful purposes, of its six member countries.[2] Both the fourth and fifth French national economic plans spelt out in considerable detail French aims and objectives in the nuclear field. The recently adopted provisions of the fifth plan provide for a total of 2,000 MW by 1970; to achieve this goal it is intended to construct at least one new 500 MW nuclear power station each year between 1966 and 1970, while the construction of three further plants of similar capacity during this period is also under study. All these plants were originally intended to burn natural uranium. This deliberate choice of the French in favour of natural uranium was due at least in part to political reasons; the main exponents of nuclear power plants using enriched uranium have been the Americans and the desire to be free from American influence both in plant design and, more particularly, for supplies of enriched uranium, has been a fundamental factor in influencing the choice of the French Government in favour of plants burning natural uranium. (But it must be noted that there have more recently been signs of a definite shift in French thinking towards the use of enriched uranium.) The French have let it be known that they consider that full competitivity with conventional power plants may be achieved by 1968[3]; elsewhere in Europe this view is, more often than not, regarded

[1] See M. Camps: *Britain and all the European Community 1955–63*. Published Oxford University Press, London, 1964, p. 55: 'M. Monnet . . . wanted to give priority to Euratom, which he felt would be negotiated and ratified . . . quickly, thus maintaining momentum in making 'Europe'. He had a number of reasons for this view. Atomic energy was a new field, so that there were, as yet, no vested interests. The French were clearly interested in some kind of European atomic energy organisation'.
Ibid. p. 68: 'The onslaught on the idea of committing France through its participation in Euratom to forgo developing atomic energy for military purposes came from all sides but particularly from the Gaullists. Here the Government offered a compromise: France would commit itself not to explode an atomic bomb before 1 January 1961, meantime there would be no bar to the research and development of the atomic bomb. This compromise was, in effect, the end of any real hope that Euratom might play a part in stopping the spread of nuclear weapons. As French Ministers who supported Euratom were at pains to point out to their critics during the debate (in the French Assembly in July 1956), the very limited French commitment had little meaning, for no one supposed that France would, in any case, be in a position to test an atomic bomb before 1961'.

[3] See in this connection an article by D. J. Rees in the *New Scientist*, no. 424 of 31 December 1964, p. 891: 'Nuclear plant studies four or five years ago showed that 250 MW to 300 MW water-type reactor plants were competitive with fossil fuel plants

Energy imports of the Community countries with and without nuclear power[1]

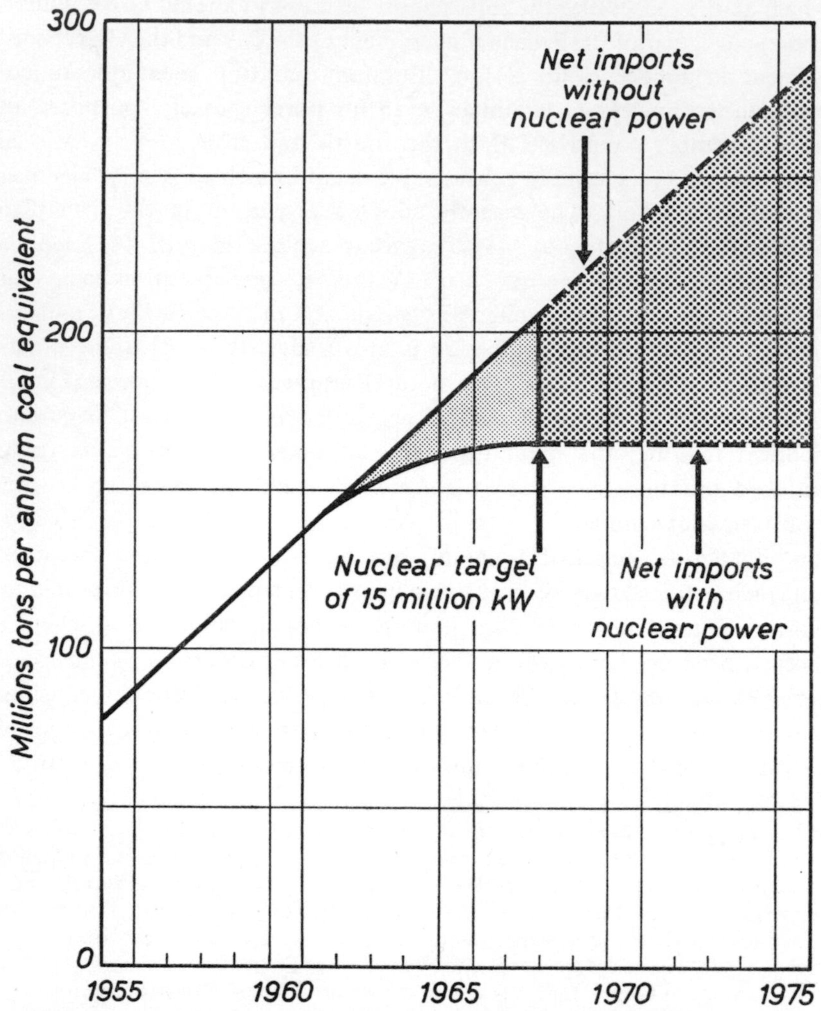

in areas where the fossil fuel costs were 35 cents per million Btu or greater. Increasing the size of units to the 500 MW to 600 MW range has now made it possible for water-type reactor plants to compete with or show an advantage over fossil fuels in areas where fossil fuel costs are in the range of 20 to 25 cents per million Btu. Unit size is an important factor for both fossil fuel plants and nuclear fuel plants, but the unit capital cost in dollars per kW of plant capacity falls more rapidly with increasing unit size in the case of nuclear plants. There are more large components such as the containment, shielding and radio-active waste treatment in nuclear plants that have disproportionate relationship with unit size than is the case for fossil plants. While the largest effect of increasing unit size is in the capital cost, there is also a small reduction in nuclear fuel-cycle cost; on the other hand, the present day mining and transport costs for coal are essentially unaffected by unit size'.

[1] Taken from *A Target for Euratom*, p. 21.

as somewhat optimistic in the light of present prices of conventional fuels and break-even point will not, it is argued, be reached before the beginning of the next decade.[1] Be that as it may, the relatively large-scale development programme of nuclear power plants, as outlined in the latest French national economic plan, may be deemed to have been motivated as much by political as by economic considerations.

Eastern Europe and the U.S.S.R.

Comparatively little information has been released about the nuclear power station programme of the U.S.S.R. According to the Russian engineer, V. S. Emelyanov, the Soviet plan in 1958 included three large stations of between 400 and 600 MW capacity then under construction at Voronesh, Beloyarsk and Ulyanovsk, as well as two others of a similar size which were planned for Moscow and Leningrad respectively. Already in operation at that time were 100 megawatts of the 600 MW plant being constructed in Siberia and the 5 MW experimental plant near Moscow, already referred to, which came into operation in 1954.[2] By 1961, however, it appeared that this comparatively ambitious programme had been cut back to four experimental plants, including two with pressurised water reactors. More recently, in 1964, in a paper given at the Third Geneva Conference[3] it was stated that the Russian authorities, while concentrating on developing the immense hydro-electric resources of Siberia, were anxious to develop breeder atomic reactors for electricity-generating purposes in European Russia. The paper went on to say that atomic power plants were still substantially more expensive than conventional plants. Nevertheless the experience obtained during the last ten years had proved immensely valuable and very real progress had been made. The answer to economic and fully competitive atomic power-stations lay in increasing their capacity, *i.e.* to match that of the largest conventional plants of 1,200–1,400 MW or even more. In the rest of Eastern Europe there are, up to the present time, only two nuclear plants under construction: the first of

[1] The likely cost of nuclear power-stations over the next few years is still very far from clear. As we have already seen, estimates vary widely and any assumptions with regard to the time at which nuclear plants will become fully competitive with conventional power-stations are still, to some degree at least, of a speculative nature. Nevertheless, although the choice of a definite date would be extremely arbitrary, it does now seem fairly certain that break-even point will be achieved somewhere between 1970 and 1975.

[2] V. S. Emelyanov: *The Future of Atomic Energy in the U.S.S.R.* Proceedings, Second Geneva Conference. Vol. I, p. 68.

[3] Paper by A. M. Nekrasav and G. S. Slevinsky: 'Progress of power engineering in the U.S.S.R. and atomic power plants'. Third Geneva Conference, 1964.

these, which is being built in Czechoslovakia, will have a net capacity of 150 MW; the second, which has been completed, is a small 70 MW experimental plant in East Germany.

Table 37. *List of nuclear reactors in operation or under construction in Europe (other than the Common Market countries and the United Kingdom) in mid-1966*[1]

Country	Site or designation	Type of reactor	Type of fuel	Capacity net MW	Date of entry into service
Czechoslovakia	Bohenice	HWGCR	UO_2	150	1968
E. Germany	Rheinsberg	PWR	UO_2	70	1966
Norway	Halden	HBWR	UO_2	20	1959
Spain	Don	HWOCR	UC	30	1968
	Zorita	PWR	UO_2	153	1967
	Sta. Maria de Gerona	BWR	UO_2	440	1969
Sweden	Agesta	PWR	UO_2	10	1963/64
	Marviken	HBWR	UO_2	193·5	1969
	Oskarhamnsverket	BWR	UO_2	400	1970
Switzerland	Lucens	HWGCR	U/G	8	1967
	Beznau	PWR	UO_2	350	1969/70
Soviet Union	Obnimsk	PWR	UO_2	5	1954
	Troitsk	PWR	U	6 × 100	1960
	Voronesh 1	PWR	UO_2	240	1964
	Voronesh 2	PWR	UO_2	360	1966
	Beloyarsk 1	BWR	U–Mo	100	1962
	Beloyarsk 2	BWR	U–Mo	200	
	Melekess	BWR	UO_2	50	
	Shevchenko	LMCR	$(Pu_1U)O_2$	350	1968/69

Northern Europe

In Northern Europe, *i.e.*, Scandinavia, where there are substantia hydro-electric resources, as well as a current plentiful availability of conventional fuels, the need for nuclear energy to supplement dwindling indigenous resources or to reduce the import bill for fuel was until recently regarded as being considerably less urgent or pressing. As a result the Scandinavian countries tended to concentrate their efforts on small research plants or experimental reactors. Thus, in Denmark, a programme with three operating research type reactors was set in motion. Norway has a 10–20 MW experimental heavy-water moderated boiling water reactor operating at Halden, primarily for process heat and is also co-operating

[1] For technical details of these reactors as well as those shown in Tables 36 and 38, see *Nucleonics*, vol. 24, no. 8 of August 1966.

with the Netherlands in establishing a joint research centre at Kjeller, near Oslo. In Sweden, the emphasis was at first put on small combined heat and power units; within the last two years, however, the Swedes, changing their line of policy, have started to build two large plants at Marviken and Oskarhamnsverket of some 200 and 400 MW respectively. The first of these plants is scheduled to come into service in 1969; the second in 1970. (Details of nuclear reactors in operation or under construction in all parts of Europe other than the Common Market countries and the United Kingdom are given in Table 37).

United Kingdom

Of all the European countries it is, however, the United Kingdom which has put in hand the largest and most ambitious nuclear power programme. Faced in the early 1950s with an anticipated fuel shortage, the British Government of the day decided to embark upon a substantial nuclear power plant building programme. The original target, announced in 1955, was for 1,500–2,000 MW of nuclear capacity by 1965. In 1957, after the Suez crisis, it was announced that this would be increased to between 5,000 and 6,000 MW by 1965. Subsequently however, this programme was twice stretched out so that by 1961 the reduced pace implied that the target of 5,000 MW would not be achieved until 1969. In April 1964 a Second Nuclear Power Programme was announced, providing for the building of an additional 5,000 MW of nuclear capacity between 1970 and 1975. Details of atomic power stations in operation or under construction in the United Kingdom by mid-1966 are given in Table 38.

By the early 1960s, as a result of longer operating experience of the experimental plants already in service, further improvements in efficiency and cost-reducing specifications arrived at by means of continuous experiments, and the successful incorporation of some new types of plant, the Secretariat of the Economic Commission for Europe in Geneva in its report on the 'Situation and Prospects of Europe's Electric Supply Industry in 1960/61' felt able to state 'that a consensus of opinion seemed to be emerging in several countries in favour of nuclear power plants becoming commercially competitive by about 1970'.[1] Reactor systems which appeared to satisfy most conditions for practical and potentially economic exploitation included two main categories: first, the gas-cooled graphite moderated systems being developed in the United Kingdom and France; second the pressurised and boiling water types favoured in the United

[1] *Situation and Prospects of Europe's Electric Power Supply Industry 1960/61:* United Nations, Geneva, 1962, p. 45.

States, the Soviet Union, Canada and Sweden. A situation will probably develop therefore, in future, in which countries, for reasons of experimental research and experience of running costs, will opt in favour of a nuclear power plant construction programme including plants with different reactor systems.[1]

Table 38. *List of nuclear reactors in operation or under construction in the United Kingdom in mid-1966*

Site or designation	Type of reactor	Type of fuel	Capacity net MW	Date of entry into service
Calder Hall	GCR	U	185	1956–59
Chapel Cross	GCR	U	190	1959
Berkeley	GCR	U	275	1962
Bradwell	GCR	U	300	1962
Dounreay	Experimental	U–Mo	15	1963
	LMCR	$(Pu_1U)O_2$	250	1971
Windscale	AGR	UO_2	28	1963
Dragon (U.K.A.E.A./ O.E.C.D.)	HTGCR	UO_2	20	1964
Hinkley Point A	GCR	U	500	1964
Hunterston	GCR	U	320	1964
Trawsfynwydd	GCR	U	500	1965
Dungeness A	GCR	U	550	1965
Sizewell	GCR	U	580	1966
Oldbury	GCR	U	600	1967
Winfrith	SGHWR	UO_2	100	1968
Wylfa	GCR	U	1,180	1969
Dungeness B	GCR	UO_2	1,200	1971

During the eight years between the end of the coal and energy shortage in 1957 and 1964, therefore, nuclear energy remained essentially in an experimental stage from which it is still only slowly emerging. The unforeseen fall in prices of imported fuels—notably oil—and the pressure this has brought about on indigenous fuels to maintain existing price levels to some extent retarded the day when nuclear energy will be fully competitive with conventional fuels. Nevertheless, for reasons of security of supply and anxiety to be fully abreast of developments in the nuclear field, as well as for political prestige, a number of European Governments embarked on substantial nuclear power station building programmes. Within the last twelve to eighteen months, moreover, there has been a further complete

[1] On this point and the optimum size of nuclear power plants see article by M. P. Ailleret: 'Integration économique des centrales nucléaires dans les ensembles de production et de distribution d'énergie électrique' in *Revue Française de l'Energie*, September 1964, pp. 577–85.

swing of the pendulum and it is now generally conceded that nuclear power will break even with conventional fuels for power-generating purposes within the next five years and may, indeed, become substantially cheaper. The question is no longer whether nuclear power will prove, economically, more attractive to the buyer than coal or oil, but simply at what exact moment in time this will happen. The only brakes on the contribution of nuclear plants in the future are likely to be their high initial capital cost and, above all, the growing difficulty of finding suitable sites.

By 1965, however, electricity produced from nuclear power stations still only accounted for 21,410 million kWh in the whole of the European O.E.C.D. area out of an overall production of electricity from all sources of energy of 771,688 million kWh. Electricity produced by nuclear power stations in 1965 amounted, therefore, to only a little over 2% of total electricity produced in the European O.E.C.D. area.

Table 39.* *Production of electricity in 1965 by source of energy in the European O.E.C.D. area (in* 10^6 *kWh)*

	Conventional thermal	Nuclear	Hydro	Total
Austria	7,340		15,170	22,510
Belgium	19,999		281	20,280
Denmark	8,055		25	8,080
France	57,250	1,400	42,350	101,000
Germany	147,200	200	14,700	162,100
Greece	3,949		854	4,803
Italy	32,770	3,900	43,430	80,100
Luxembourg	1,430		986	2,416
Netherlands	23,400			23,400
Norway	50		46,900	46,950
Sweden	3,190	10	44,600	47,800
Switzerland	350		25,730	25,080
Total	304,983	5,510	235,026	545,519
United Kingdom	161,500	15,900	4,400	181,800

* *Source:* O.E.C.D. *15th Electricity Industry Survey*. Published Paris, 1965. Figures for Iceland, Ireland and Turkey have not been included in this table.

The exploitation of natural gas in Europe

Natural gas (as a major form of energy) has, like nuclear power, been a recent addition to Europe's energy supplies. Although the existence of combustible gaseous emanations had been known in various parts of the continent, notably in Italy,[1] for nearly two thousand years, it was not

[1] In Tuscany. For a totally different slant see H. V. Morton *A Traveller in Italy*, p. 261 (ENI plant near Ravenna) and reference to the *Waters of Clitumnus*, pp. 607–10.

until 1869/70 that the first small industrial application took place. This occurred at Salsomaggiore near Piacenza, in the Po valley, when it was used to recover salt through the evaporation of brackish waters. Subsequent progress in the development of natural gas deposits was however extremely slow and, by 1939, annual natural gas production in the whole of Western Europe amounted to no more than some 50 million cubic metres. Production rose rapidly after the war, with the discovery and development of natural gas fields in Austria, Italy and France, from a level of 400 million m³ in 1946 to 5,025 million m³ in 1955. By 1959 production had again doubled and there was a further increase in output of well over 60% between 1959 and 1962. By 1964 production of natural gas had reached a total of nearly 17,000 million m³. During this latter period extensive drilling programmes were under way in a number of European countries and in 1961 the big gas field of Slochteren in the Groningen Province of North-Eastern Holland first came into operation.

Table 40. *Production of natural gas in the European area of O.E.C.D. 1955–1964 (in million cubic metres)*

	1955	1956	1957	1958	1959	1960	1961	1962	1963	1964
Austria	748	745	759	820	1,128	1,468	1,556	1,635	1,704	1,764
France	278	319	561	1,054	1,662	2,845	4,014	4,740	4,860	5,088
W. Germany	240	367	357	344	549	643	736	917	914	1,457
Italy	3,627	4,465	4,993	5,176	6,118	6,849	6,849	7,151	7,260	7,668
Netherlands	132	149	142	188	289	384	363	405	568	766
Yugoslavia	55	68	41	46	50	53	55	59	171	274
United Kingdom	—	—	—	—	1	1	3	4	4	4
Total	5,080	6,113	6,853	7,628	9,797	12,243	13,576	14,911	15,501	17,021

Until 1964 consumption of natural gas was limited to the amount that could be produced in Europe itself since facilities for importing natural gas from North Africa or North America did not yet exist. The discovery of massive reserves of natural gas in the Sahara and the construction of pipelines from the gasfields at Hassi R'Mel to the Algiers and Oran regions, capable of transporting 3,000 million m³ a year at full capacity, had, however, given a powerful impulse to research into means of commercial transportiaton of natural gas to Europe. Two main possibilities were studied: first, liquefying the gas and transporting it by tanker to various points of demand and, second, the construction of submarine gas pipelines linking the North African coast to the Spanish coast, Sicily or the

Italian mainland.[1] Up to the present time it is the former of these two methods of transporting natural gas which has proved the more economic and practical. The voyages made in 1959 by the *Methane Pioneer* between the Western Hemisphere and the United Kingdom had demonstrated conclusively that transport of liquid methane by sea was technically feasible, but translating this experience into large-scale operation raised a number of technical and economic problems. It was not until the end of 1964 that the first regular shipments of natural gas from Arzew, on the Algerian coast, to Canvey Island in the Thames estuary got under way although this was then quickly followed by the inauguration of the Arzew–Le Havre link by the French methane carrier *Jules Verne*, designed to make 33 round trips a year, during which she transports some 450 million m^3 of liquefied gas (*i.e.* approximately 415,000 tons of coal). During the whole of this period studies on the technical and economic feasibility of under-water pipelines were being continued but, so far, it has not proved possible to overcome all of the numerous technical problems involved. Some of the initial urgency has however been lost as a result of the favourable operating costs obtained by means of tanker transportation of natural gas.

The Robinson Commission in their report in 1960 estimated that production of natural gas in the O.E.C.D. European area would rise from about 5,000 million m^3 in 1955 to 20,000 million in 1965 and 40/45,000 million m^3 by 1975. This estimate was however made before the full extent of the Groningen gas field was known and well before the days of the rush for exploitation rights for natural gas under the North Sea. By 1964 the O.E.C.D. Oil Committee in their report *Oil today* stated: 'The full future extent of the indigenous resources of natural gas that may be available from outside the European area cannot be predicted with any degree of precision. However, it is certain that with the resources already discovered within the O.E.C.D. European area, natural gas will in future play a relatively more important part than it has in the past'.[2] The 1966

[1] Studies were also made of the possibility of constructing submarine pipelines from the Moroccan coast to Gibraltar and from Mostaganem, on the Algerian coast, to Cartagena. The Gibraltar route is only 15 kms. long and does not go deeper than 450 metres. The Mostaganem–Cartagena route on the other hand, while 180 km and with the sea-bed going down in places to about 2,700 metres, would considerably reduce the overall length of the pipeline from the gas fields to the consuming sectors.

[2] A similar tone of moderated optimism had been expressed by the O.E.C.D. Oil Committee in its report published in October 1962 and entitled *The search for and exploitation of crude oil and natural gas in the European area of the O.E.C.D.*, p. 153: 'Forecasts concerning natural gas seem an even thornier proposition, experience having shown the existence of huge pools like Lacq and Slochteren and the sharp impact they have on production. As natural gas is made up of hydrocarbons which

O.E.C.D. Energy Committee's Report estimated the reserves for natural gas in the O.E.C.D. European area at 2,200 thousand million m³ (some 2,850 million tons of coal). This, at current 1964 levels of production, would mean a theoretical life of about 100 years. But production rates are certain to increase, particularly in the Netherlands, while at the same time the full extent of the reserves, both under the North Sea and elsewhere, is still largely a matter for guesswork.

In October 1963 the Dutch Authorities announced that they estimated the value of the Groningen reserves at 1,100 milliard m³ or about 1,210 million tons of coal equivalent. This meant that the Groningen field was one of the largest in the world, far exceeding those of Hassi R'Mel in Algeria, with 990 milliard m³, and Lacq, with 200 milliard m³. This figure was increased in 1965 to 1,700 milliard m³. It is, however, widely believed that the real figure for the Dutch reserves is very much higher than the official figure announced by the Dutch Government, which is a highly conservative estimate of the Slochteren reserves only.[1] The Slochteren deposits together with those reputed to lie off the north coast of Holland and the north-west coast of the Federal German Republic[2] are con-

are chemically most stable, it is less sensitive to geological disturbances and more proof against pressure and temperature stresses. At great depths and in certain basins where the sedimentation is of less distinctive marine origin, the chances of finding gas are known to be better than those of finding liquid hydro-carbons. On these grounds, there is reason to believe that the European prospects for natural gas are still very good. In the light of recent discoveries in the Netherlands, it is possible to foresee a substantial increase in production which might amount to between 400 and 500 million m³ in 1965'.

[1] One important factor that should be borne in mind is that the geological reserves can never be recovered in full. The portion of the total amount of natural gas that is recoverable is linked closely to the period of time over which the field is exploited. If the gas is drawn off over a period of some 20 to 30 years, then some 60–70 % of the gas may be recovered; if, on the other hand, the deposits are exploited over a short period of time, say ten years, then the total amount recoverable is unlikely to exceed 40 %. These factors were taken into account by the Dutch authorities when they made their statement towards the end of 1963 about the size of the Slochteren reserves: the figure of 1,100 milliard m³ relates to recoverable resources of natural gas over a period of some 30 years. The same considerations naturally apply to the higher 1965 estimate.

[2] The Ems–Dollart offshore area, found by the mouth of the river Ems, where both the Netherlands and Western Germany adjoin the sea, has recently been the subject of an agreement between the two countries, whereby all exploration and exploitation of hydro-carbons will be the joint concern of concession holders on each side of the border. It was announced in October 1963 that gas had been found in substantial quantities near the town of Bierum on the Dutch side in a well which N.A.M. had been drilling. In accordance with the agreement, the German registered Gewerkschaft Brigitta is participating financially in this project and is entitled to a half-share in the gas. Like N.A.M. this company is jointly owned by Shell and Esso.

fidently believed by geologists to constitute a part of a very much larger gas field extending well into the North Sea. It is this widely-held assumption —based on preliminary boring and test-hole results and backed by favourable geological conditions—which have led to the large-scale exploration and drilling programmes now proposed for large areas of the North Sea. Under the Geneva Convention, signed in 1958, the mineral rights under the waters of the North Sea were divided among the countries bordering upon it, *i.e.*, Belgium, Denmark, Germany, Netherlands, Norway and the United Kingdom, with the last mentioned getting the lion's share. While work off the Dutch and German coasts has been in progress for some considerable time, drilling in the North Sea, within the areas allocated to the United Kingdom under the Geneva Convention, began in earnest in the spring and summer of 1965. The closing date for licences for search for and exploitation of natural gas in the North Sea was 20 July 1964; by that date, the Ministry of Power in London had received applications from 31 groups or consortia, consisting mainly of oil companies, covering a total area of 95,000 km^2—the total United Kingdom share of the North Sea continental shelf amounts to some 250,000 km^2, which has been divided into sections of 250 km^2. These licences, which were granted in November 1963, confer the exclusive right of exploration for and production of oil and natural gas for a period of six years with an option for a maximum of half the area covered in the initial licence for a further 40 years. The cost per licence was £25 per km^2 for the first six years, rising thereafter to £40 a year and then by annual increments of £25 to a total of £290 per km^2. In the event of oil or natural gas being discovered royalty payments are required amounting to 12·5% of the value of the oil or gas at the well-head.

At the time of writing (in October 1966), 21 wells had been drilled in the British area of the North Sea. Of these 14 had been unsuccessful, while the other 7 have had varying degrees of success. Judging by the figures released by the oil companies and their associates, the value of all finds up to that time probably amounts to some 24 milliard m^3 (*i.e.* some 22 million tons of coal). Although natural gas from the B.P. well will be flowing into British homes by the beginning of 1967, the real impact of natural gas is hardly likely to be felt until the early or middle 1970s. But the big all-important question which has not so far been answered—and it is vital both with regard to the quantity of natural gas that is likely to become available and the price at which it will be sold—is whether there is one vast North Sea gasfield or whether natural gas deposits are divided among a number of pockets of limited size.

In view of the current interest in natural gas we have briefly examined below the development of the natural gas industry in the countries with the largest production and reserves of natural gas at the present time.

Italy

As we have already seen, the first industrial application of natural gas in Europe occurred in Italy at Salsomaggiore in 1869/70. By 1890, a natural gas distribution network for both domestic and industrial purposes for the whole Salsomaggiore area had been completed. Methane issuing from freshwater wells was also being used quite extensively at about this time in various places in the Po delta, notably at Polesine, for private domestic purposes. The first systematic development of the industrial utilisation of methane however did not take place until 1932/33, again at Polesine, where exploitation was begun of relatively superficial fields at depths down to 700 m where the methane was normally found in solution with brackish water. The gas obtained in this way was used as a fuel. Often, however, in these cases, the recovery of the gas was made more complicated by the need to dispose of the brackish water without endangering the whole operation. In some cases the volume of the water was equal to the volume of the gas recovered. In contrast to these earlier efforts, the immediate post-war exploitation was of deeper fields and the authority responsible for their development, the Azienta Generale Italiana Petroli (AGIP) greatly increased the output of natural gas, from a level of 200 million m^3 in 1938 to some 1,600 million m^3 by 1952. The utilisation of methane by the Italian gas industry has been a very gradual process, progressing comparatively slowly through a number of stages, that is, methane enrichment of manufactured gas, re-forming of methane, utilisation of air-methane mixtures, supply of straight natural gas through existing distribution networks. By 1964 production of natural gas in Italy had reached a total of 7,668 million m^3.[1] A network of pipelines has been built, the greater part of which lies in the Po valley which includes the Cortemaggiore–Turin (192 km), Cremona–Venice (190 km) and Cortemaggiore (142 km) sections, all of which have a 16 inch diameter.

Although it is generally considered that, as a result of their intensive exploitation, production from the Po valley gasfields is likely to decline in

[1] The major area of production remains the Po valley; the State corporation, Ente Nazionali Idrocarburi, has a monopoly. The Romagna accounted for 5,600 million m^3 of the 1963 total, and Lombardy for 1,400 million m^3. Production outside the Po valley included 77 million m^3 from Abruzzimilise and 58 million m^3 from Sicily.

future, drilling and boring programmes are in full swing both in central and southern Italy and in Sicily. Furthermore, given the extensive natural gas pipeline network that already exists, the Italians are now seeking to augment their indigenous supplies by importing natural gas from other countries. Agreement has already been reached in principle upon the import of 3 milliard m^3 from Libya via Genoa, while negotiations are currently also in progress with the Russians (see final short section on Eastern European countries).

France

French discoveries of natural gas are mostly the result of post-war exploration although the Société Nationale des Pétroles d'Aquitaine (SNPA) was created in 1941 with a majority holding of the capital being retained by the then French Government. At that time it was known that a small gas field existed at St Marcet, some 65 km south of Toulouse and the organisation took its name from the French province of Aquitaine of which Toulouse was the capital. Small amounts of natural gas were obtained from the St Marcet field as early as 1942 but production did not really get under way until after the war.

The big discovery of natural gas in Provence was made in 1951 at Lacq, then a small village, some 120 km west of St Marcet on a site where a small quantity of oil had previously been discovered at a depth of some 700 m. The gas was struck at a depth of well over 3,000 m, and was found to be of very high pressure and with an exceptionally high sulphur content. The well got out of control and burnt in a gigantic plume of flame for two months before it could be extinguished.[1] It was always evident that sulphur recovery would be a major element in the successful utilisation of the Lacq gas and it was largely as a result of this particular problem that full-scale production was delayed until April 1957 when the first section of the Lacq plant came into operation with an output of some 8 million m^3 of crude gas a day. By 1963 the volume of annual natural gas production was 7,518 million m^3. About 18% of the total gas sold is disposed of to industries located in the vicinity of Lacq (including one large power-station belonging to Electricité de France, which takes about 8 million m^3 of gas a day) through a pipeline distribution network with a total length of 5,843 km. Two separate companies were created for the supply of natural gas to the south-west of France and the remainder of France respectively. The

[1] To do this the French called upon a well-known American authority on the extinction of oil fires, N. Myron Kinley—see article in *Coke and Gas*, of March and April 1960, by L. T. Minchin: 'The Lacq Gas Field, a Revolution in French Gas Practice'.

company which controls supplies of natural gas to all parts of France other than the Lacq region and the south-west is the Compagnie Française du Methane, of which half the capital is held by SNPA and half by Gaz de France; while gas is supplied from SNPA's huge underground storage plant at Lussagnet through a distribution network of nearly 4,200 km in length.

Recently, the SNPA is reported to have made a further discovery of natural gas at a location some thirty or forty miles from Lacq. The value of this find is however understood to be no more than one-fifth of the major Lacq field. Further boring programmes are still in progress. Drilling for gas is also in progress in the Bay of Biscay.

Like the Italians, the French have also turned to outside suppliers and, in addition to their imports from Algeria, have recently negotiated a contract for an annual level of imports of 5 milliard m³ from the Netherlands.

Netherlands

Royal Dutch Shell began searching for petroleum in Holland under an exploration licence in 1935 and the first wells were drilled during the period 1937–39. Their first success was achieved at Coeverden in the eastern part of the country in 1936. Subsequently, substantial crude oil deposits were located at Schoonebeek, which later became the largest oil-field in Western Europe. Promising signs of natural gas were first found in the province of Overyssel and other areas in February 1943. Although these were not in themselves exploitable they were of sufficient interest to justify the drilling of test wells specifically for natural gas. After the war, Esso started prospecting and, in 1947, the two groups formed a joint company, Nederlandse Aardolie Maatschappij, or NAM, and increased the scope and extent of their explorations. The first deposit of pure natural gas to be found was in June 1948 near Coevorden. Other minor finds followed but it was not until the early part of 1960 that the big discovery of the Slochteren deposits was made.

Public attention was first drawn to the importance of the finds in Northern Holland as a result of a debate on energy in the European Parliamentary Assembly in Strasbourg, when it was suggested that the extent of the natural gas deposits that had been discovered could be estimated at 300 milliard m³. After some initial statements which tended to play down the value of the reserves, the Dutch Government announced, early in 1961, that the extent of the proven reserves was estimated at 150 milliard m³; by the beginning of 1963 the figure had gone up to 450 milliard m³, and in October 1963 the Dutch authorities announced that they estimated the value of the reserves at 1,100 milliard m³. Later, in November 1964, the Dutch Minister for Economic Affairs, replying to a question in the Dutch Parliament by one of the Deputies,[1] after denying speculations about a true reserve figure of some 3,000–5,000 m³, stated that work was at present in progress on a new assessment of the reserves. More recently, in 1965, a statement was made showing a reserve figure of some 1,700 milliard m³.

Under the terms of a contract signed in 1954 between the Dutch Government and NAM, the distribution of natural gas was left in the hands of the State. Payment to the company was based on an agreed price and on the availability of gas at the well-head, not on actual off-take. While this scheme worked well enough for the limited quantity of gas available from the fields in East Holland and elsewhere, it became completely inoperable in the presence of the large reserves discovered in Groningen. The new organisation which was set up to control the exploitation of the natural gas is a complex one. Under it NAM retained the rights of physical exploitation of the concession at Groningen, but all the gas produced is sold to a new company, set up in 1963, called Gasunie, in which the State, Shell, Esso and the Dutch State Mines participate. The State has 10 %, Shell and Esso both have 30 % and the Dutch State Mines 40 %. The company also pays 10 % of its profits in royalties to the Government. With regard to management policy, the State Mines have 50 % of the voting power and Shell and Esso 25 % each. Gasunie took over the existing transmission network of the State Gas Board and the State Mines and for the first two or three years also co-ordinated the distribution networks for blast-furnace gas in Northern Holland. In 1966, however, Gasunie bought out all the Hoogovens (*i.e.*, Dutch Steelworks) sales contracts and pipeline network for a lump payment of some £18 million. The Government retained power to control the wholesale prices charged for the gas, to approve the purposes for which gas is sold and investments in the transmission grid, and to determine any special sales contracts that may be made to promote particular industrial developments. The inland use of Groningen gas was concentrated initially on domestic town gas for cooking and water heating appliances and for certain specific industrial uses where the greatest advantage can be obtained from its special properties. It was however planned to extend its use for space heating by stoves and central heating

[1] See also question put in the Dutch Lower House (Tweede Kamer) by MM. Blaisse and Maenen on 1 December 1964 and reply given by the Dutch Minister of Economic Affairs, Dr J. Andriessen, on 11 December 1964.

and to the power stations.[1] Since then, however, it has become clear that the domestic and general industrial sectors alone will not be able to take all the natural gas available for marketing in Holland. The latest estimates suggest that natural gas availability by the early 1970s will not be less than 35–45 milliard m^3 a year and although some 23–25 milliard m^3 have already been sold on the export market, this still leaves some 12–20 milliard m^3 for indigenous consumption. In fact, it is known that the Dutch Government have intimated to the power stations that they should count on taking about one-third of their energy requirements by 1970 in the form of indigenous natural gas. In order to distribute the natural gas, a vast network of pipelines is being built. Thus by the end of 1965, some 700 km of main lines and nearly 2,000 km of subsidiary pipes had been laid. This was, of course, in addition to the existing gas network taken over by Gasunie, totalling some 3,000 km, which also carries natural gas. Recently, Gasunie have announced their intention of building a second main pipeline from Slochteren to the industrial west and centre of Holland (and for possible additional export contracts). With a diameter of 120 cm this will be the largest pipeline yet in Europe.

The main difference between the Dutch and the French and Italian natural gas fields—apart from the obvious fact that it is a much larger deposit—lies in the fact that approximately half the amount to be produced by 1970 is intended for export. A special export organisation, NAM Gas Export, has been set up to promote sales of natural gas abroad. With the Groningen area so close to the Netherlands north-east frontier, several major potential domestic and industrial centres of consumption in Belgium, the Ruhr and Northern France are within a short distance.

Up to the present time, contracts have been negotiated with consumers in Belgium (5 milliard m^3), France (5 milliard m^3) and Germany (13–15 milliard m^3) for a total quantity of some 23–25 milliard m^3.

[1] In a Memorandum on natural gas, dated 11 July 1962 addressed to the Second Chamber of the Dutch Parliament, the then Minister for Economic Affairs, M. de Pous, stated that he thought the natural gas would be mainly used in the public gas supply system. He did not regard industry and electric power-stations as likely outlets for natural gas. It was important to use the natural gas in those sectors of the market where the advantage to the country's economy would be the greatest possible. It might therefore be considerably more advantageous to export the natural gas and to continue to import cheap oil or coal for industry and the power-stations. Since this statement by M. de Pous however the volume of proved reserves has risen substantially. On the assumption that the objective of the Dutch Government is the production of some 40–45 million tons of coal equivalent by 1970, including some 20–25 million tons for export—and all the evidence points to the fact that it does—then it is questionable whether the domestic and general industrial sectors, whether through conversion or additional demands for energy, will be able to account for the remainder.

Before leaving the subject of natural gas in Western Europe, it is worth outlining the differences that exist, despite the relatively short period of time that has elapsed since the first important gas finds occurred, in the organisation and marketing policies and objectives in the different Community countries. In Germany, as one would expect, sales of natural gas are largely in the hands of the oil companies. In France and Italy State control is complete. In Holland, although marketing as such is largely left to the oil companies, State influence on policy is paramount. But it is even more important to see the influence of natural gas in each country in the right perspective. In France, Germany and Italy natural gas, while increasing absolutely in terms of the level of consumption, is likely to remain of secondary importance. In the Netherlands, on the other hand, natural gas is clearly destined to become one of the main sources of energy. The Dutch Government has from the outset set two main objectives for its marketing policy. The first and more important of these has been the promotion of the use of natural gas with a view to meeting about one-third of the country's energy requirements from this source by the beginning of the next decade. At the same time, they have shown their anxiety to make the fullest possible use of the specific qualities of natural gas by encouraging and stimulating its use in the first instance by industrial and domestic consumers. Supplies to power-stations are likely to be limited to one-third of the total energy requirements of the electricity sector. The second objective has been to secure export orders for Dutch natural gas—a field where, as we have already seen, they have had considerable success. In France and Italy the local importance of natural gas will be enhanced, however, by its use as a factor for the promotion of regional development.

Eastern Europe and the U.S.S.R.

According to some recent reports, there is a possibility that the Soviet Union may shortly plan to extend its natural gas pipelines to the borders of Western Europe. A pipeline to Finland has also been suggested. The latter would require only a spur of some 400 km from Leningrad and would be able to supply Finland with 80% of her gas requirements by 1970. Production of natural gas in the Soviet Union in 1964 amounted to over 125 milliard m^3 and is scheduled to rise to between 310–325 milliard m^3 by 1970. Proved reserves of natural gas at the end of 1963 were estimated at 2,786 milliard m^3 and potential reserves at 60,000 milliard m^3. Transmission pipelines for natural gas in the Soviet Union in 1965 totalled nearly 50,000 km with a further 12,000 km either under construction or in the planning stage. Most of this network has in fact been constructed during the last seven or eight years. It is planned to extend the network still further in future and forward projections include one very big new pipeline from the gasfields of Bukhara to the northern part of European Russia and the Baltic States. The Russians are also going ahead with the construction of a large number of gas storage reservoirs. These include several big underground reservoirs in the Moscow area, as well as plans to use the old Armenian saltmines for storage purposes.

It is understood that the Russians are currently negotiating with the E.N.I. and the Italian Government for a total annual export of the order of 6 milliard m^3 for delivery to northern Italy. A special feature of the negotiations—as well as a strong inducement to the Italians—is the Russian proposal that they would be prepared to purchase the steel pipes for the pipeline from the Ukraine to Italy from the Italian State steel company, Finsider. The project is however being strongly resisted in some high-level circles in Italy on the grounds that it would result in an excessive degree of reliance upon natural gas imports from a Communist state.

Other major importers of natural gas from the Ukraine include Poland, which takes 5 milliard m^3 a year, and Czechoslovakia with 3 milliard m^3 a year.

Large natural gas deposits have recently been discovered in Poland itself, but have not so far reached the development stage. New pipelines are under construction in almost all the eastern European countries, both in order to develop new indigenous fields, as is the case in Poland and Hungary, and to import natural gas from the Soviet Union. It is clear, therefore, that in Eastern Europe, no less than in the West, natural gas is fast becoming an increasingly important source of energy. It is towards its potential contribution, together with that from the other major sources of energy, that we shall turn in Section III.

Internal consumption of commercial sources of primary energy in Europe, 1950 and 1955–65 (in '000 tons of coal equivalent)

Solid fuels

	1950	1955	1956*	1957	1958*	1959	1960	1961	1962	1963	1964	1965
Western Europe												
Austria	7,634	8,418	8,783	9,181	8,154	7,486	7,987	7,482	7,863	8,571	8,143	7,711
Belgium	25,469	27,953	28,542	27,624	23,339	24,274	24,363	24,370	25,909	27,106	24,980	23,941
Denmark	6,581	6,952	6,197	4,906	4,650	4,540	5,688	5,400	5,867	6,042	5,767	5,143
Finland	1,927	2,425	2,739	2,792	2,595	2,822	3,146	3,012	2,946	2,508	3,109	3,266
France	62,850	69,386	77,464	80,039	73,234	69,104	68,582	69,586	71,390	72,316	73,341	76,301
West Germany	116,361	151,751	162,909	164,929	153,953	141,727	159,187	154,882	160,131	163,713	157,709	148,035
Saar	6,701	8,692	9,347	9,403	9,265	9,110						
Greece	440	603	673	726	710	834	1,260	1,394	1,463	1,801	1,981	2,458
Italy	10,199	11,748	12,523	13,210	10,796	10,481	11,537	12,054	12,596	13,145	11,595	11,978
Ireland	2,275	2,169	1,608	1,529	1,594	1,801	1,886	1,977	1,681	1,654	1,530	1,465
Luxembourg	2,988	3,934	3,997	4,175	3,582	3,695	4,069	4,046	3,869	3,641	3,824	3,623
Netherlands	16,384	17,406	18,650	18,229	16,437	15,389	16,053	15,698	16,526	17,549	16,099	13,936
Norway	1,839	1,498	1,678	1,261	1,140	1,079	1,122	1,052	1,012	1,241	1,103	1,207
Portugal	1,147	859	965	1,134	1,025	920	902	1,099	1,010	1,154	1,115	1,084
Spain	12,654	14,057	14,190	15,990	16,860	14,783	15,222	15,701	15,978	15,902	14,786	15,945
Sweden	7,442	6,003	5,368	5,156	3,458	3,385	3,974	3,542	3,386	3,545	3,564	3,010
Switzerland	2,468	2,756	3,162	3,325	2,342	2,280	2,595	2,256	2,319	2,823	1,979	1,671
Turkey	3,043	4,319	4,512	5,002	4,947	4,688	4,723	4,351	4,873	5,289	5,794	5,703
Yugoslavia	7,025	9,211	10,549	11,257	11,202	12,006	13,545	14,238	14,439	15,880	17,009	17,960
Total	295,432	350,140	373,856	379,861	349,283	330,404	345,841	342,140	393,258	363,880	353,438	335,437
United Kingdom	203,058	217,083	218,421	212,876	201,749	187,488	196,746	194,486	190,375	195,117	189,016	185,063
Eastern Europe												
Bulgaria	2,895	4,916	5,141	6,096	6,529	7,871	9,104	9,675	10,994	11,550	12,699	13,000
Czechoslovakia	31,568	42,199	48,554	49,286	53,264	52,188	54,223	58,473	61,367	64,191	65,484	65,000
East Germany	49,164	69,499	70,485	72,366	73,591	73,281	77,840	81,271	84,864	88,279	88,090	86,530
Hungary	8,258	12,625	11,691	13,487	13,797	14,450	15,404	16,524	16,765	18,695	20,285	19,347
Poland	50,945	69,061	75,250	79,645	78,529	83,560	87,553	90,159	93,481	98,956	102,252	102,154
Rumania	2,098	4,290	3,006	3,598	3,700	3,822	4,570	4,800	8,849	9,479	10,020	10,716
Total	144,568	202,590	214,127	224,478	229,770	235,172	248,694	260,902	276,320	291,150	298,830	296,747
U.S.S.R.	205,700	336,851	325,100	351,100	362,700	370,000	432,952	429,145	432,593	444,333	454,000	481,900

* Unrevised figures

Source: United Nations Statistics.

Internal consumption of commercial sources of primary energy in Europe, 1950 and 1955-65 (in '000 tons of coal equivalent)

Liquid fuels

	1950	1955	1956*	1957	1958*	1959	1960	1961	1962	1963	1964	1965
Western Europe												
Austria	818	2,229	2,595	2,745	3,055	3,312	3,900	4,346	5,271	6,068	6,939	7,917
Belgium	2,823	5,183	6,810	6,700	7,715	8,373	9,329	10,074	12,003	13,380	16,226	19,863
Denmark	2,301	4,244	5,130	4,695	5,405	6,188	7,002	7,946	9,270	10,794	12,046	14,799
Finland	761	1,695	2,018	2,747	2,700	2,625	3,465	3,810	4,500	5,250	5,800	6,250
France	13,811	21,923	25,980	25,290	28,530	28,520	31,755	34,700	41,312	49,050	57,792	70,350
West Germany	5,171	14,747	17,265	19,770	25,350	30,845	38,907	48,600	69,492	82,860	96,328	104,328
Saar	72	161	165	180	225	263						
Greece	1,356	1,743	2,010	1,975	2,135	2,406	2,877	3,150	3,455	3,519	4,344	4,704
Italy	6,065	13,670	16,470	17,360	19,320	21,155	25,780	30,551	38,529	44,670	53,600	61,911
Ireland	777	1,193	1,860	1,350	1,400	1,488	1,456	1,710	2,135	2,381	2,757	3,243
Luxembourg	81	174	210	210	255	276	315	378	588	848	1,023	1,245
Netherlands	3,504	5,891	8,475	9,025	10,345	10,463	12,288	13,656	16,173	19,000	21,606	27,161
Norway	1,926	3,432	4,062	3,464	3,987	4,308	4,926	5,025	5,091	5,727	5,878	6,357
Portugal	810	1,278	1,545	1,615	1,740	1,605	1,787	1,905	2,205	2,516	2,420	3,069
Spain	1,538	3,285	3,956	4,860	5,654	7,590	7,770	7,800	9,999	10,592	14,360	16,910
Sweden	5,079	11,028	13,155	12,410	13,825	15,134	17,086	17,027	18,584	20,943	22,326	26,589
Switzerland	1,494	2,801	3,720	3,860	4,500	4,650	5,435	6,176	7,343	9,312	9,942	12,651
Turkey	785	1,671	1,770	1,935	2,380	1,905	2,175	2,400	3,069	3,606	4,153	4,500
Yugoslavia	653	963	1,083	1,248	1,437	1,515	1,770	1,950	2,325	2,570	3,000	3,200
Total	49,825	97,311	118,279	121,439	139,958	152,621	178,023	210,204	251,248	293,086	340,080	395,047
United Kingdom	20,012	28,234	33,765	32,520	40,765	45,207	53,228	57,711	65,525	71,232	78,750	96,299
Eastern Europe												
Bulgaria	207	510	746	951	921	1,050	1,150	1,250	2,600	3,200	3,600	3,800
Czechoslovakia	549	1,500	1,764	2,414	2,724	3,150	3,500	3,850	5,200	6,200	7,000	7,400
East Germany	203	840	587	755	1,017	1,280	1,500	1,750	3,500	4,000	4,500	4,800
Hungary	543	1,760	1,530	1,924	1,890	2,216	2,446	2,801	2,962	3,386	3,930	4,268
Poland	696	1,517	2,093	2,439	2,685	3,210	3,710	4,400	4,880	5,493	6,180	7,112
Rumania	3,630	4,059	5,991	6,278	6,506	5,948	6,200	5,600	7,000	7,200	7,500	7,800
Total	5,828	10,186	12,711	14,761	15,743	16,854	18,506	20,551	26,142	29,479	32,710	35,180
U.S.S.R.	54,200	89,535	119,800	140,600	161,900	139,110	153,705	...	188,570	207,664	221,715	...

* Unrevised figures.

Source: United Nations Statistics.

Internal consumption of commercial sources of primary energy in Europe, 1950 and 1955–65 (in '000 tons of coal equivalent)

Hydro-electricity

	1950	1955	1956*	1957	1958*	1959	1960	1961	1962§	1963§	1964§	1965§
Western Europe												
Austria	1,784	2,853	3,588	3,713	3,701	3,617	3,988	3,886	4,011	4,782	5,270	6,418
Belgium	25	52	75	69	79	40	69	75	66	56	64	109
Denmark	14	12	10	12	12	10	10	10	10	10	10	10
Finland	1,460	2,476	2,081	2,648	2,736	2,224	2,112	3,096	3,812	3,246	3,400	3,795
France	6,475	10,228	10,392	9,942	12,930	13,080	16,220	15,380	14,544	17,445	13,832	18,533
W. Germany	3,454	4,801	5,110	4,851	5,195	4,365}	5,127	5,112	5,827	5,507	4,858	6,103
Saar	9	8	10	9	11	8}						
Greece	9	133	212	140	180	172	186	221	245	322	300	304
Italy	9,153	13,064†	13,239	13,464	15,153	16,180	19,284	17,620	16,641‡	18,400‡	16,381‡	19,628‡
Ireland	187	192	273	278	382	296	370	291	260	259	278	342
Luxembourg	1	1	1	1	1	1	8	22	19	195	322	367
Netherlands	—	—	—	—	—	—	—	—	—	—	—	—
Norway	7,082	8,982	9,411	10,272	10,912	11,358	12,444	13,432	15,014	15,991	17,554	19,472
Portugal	175	690	813	729	994	1,139	1,235	1,363	1,398	1,595	1,750	1,593
Spain	2,032	3,612	4,473	3,810	4,474	5,620	6,181	6,300	6,384	8,370	8,189	8,400
Sweden	6,935	8,660	9,590	10,845	11,430	11,450	12,436	14,650	15,632	15,131	17,209	18,572
Switzerland	3,894	5,599	5,707	6,023	6,234	6,464	7,004	7,480	7,802	8,671	8,842	9,919
Turkey	12	36	65	124	261	274	395	505	441	833	661	868
Yugoslavia	470	1,044	1,141	1,400	1,708	1,874	2,383	2,252	2,740	3,218	3,030	3,578
Total	43,171	62,443	66,191	68,330	76,393	78,172	89,452	91,695	94,846	104,031	101,950	118,011
United Kingdom	592	678	907	1,098	1,080	1,079	1,249	1,537	2,957	3,838	4,647	7,463
Eastern Europe												
Bulgaria	122	258	302	330	382	442	754	720	676	836	1,000	1,200
Czechoslovakia	350	770	761	834	1,040	810	992	916	1,196	910	1,090	1,782
E. Germany	92	158	185	191	200	214	247	270	244	219	240	260
Hungary	16	18	14	16	19	31	36	32	32	32	40	30
Poland	148	284	254	228	303	219	261	246	307	265	288	362
Rumania	68	129	115	120	112	119	159	186	261	215	237	404
Total	796	1,617	1,631	1,719	2,056	1,835	2,449	2,370	2,716	2,477	2,895	4,038
U.S.S.R.	5,100	9,300	11,594	15,772	18,592	19,052	20,365	23,649	28,778	30,343	31,800	—

* Unrevised figures.

† Including Geothermic Electricity amounting to 744 t.c.e.

‡ Including Geothermic Electricity amounting to:

1962	1963	1964	1965
938	967	1,011	1,027

§ Including Nuclear Energy amounting to:

	1962	1963	1964	1965
Belgium	1	17	19	—
France	169	168	231	359
Italy	—	129	957	1,425
W. Germany	39	22	40	45
United Kingdom	1,391	2,380	3,040	5,615

Source: United Nations Statistics.

Internal consumption of commercial sources of primary energy in Europe, 1950 and 1955–65 (in '000 tons of coal equivalent)

Natural gas

	1950	1955	1956*	1957	1958*	1959	1960	1961	1962	1963	1964	1965
Western Europe												
Austria	649	710	1,027	867	1,130	1,543	1,985	2,100	2,219	2,294	2,442	2,298
Belgium	20	88	117	122	125	103	85	85	87	84	80	93
Denmark	—	—	—	—	—	—	—	—	—	—	—	—
Finland	—	—	—	—	—	—	—	—	—	—	—	—
France	322	380	356	584	842	1,815	3,792	5,302	6,338	6,516	6,806	6,793
W. Germany	78	501	776	714	768	947}	1,050	1,182	1,238	1,693	2,519	3,667
Saar	—	—	—	—	—	—						
Greece	—	—	—	—	—	—	—	—	—	—	—	—
Italy	667	4,703	5,801	6,450	6,735	7,920	8,362	8,894	9,267	9,335	9,800	10,391
Ireland	—	—	—	—	—	—	—	—	—	—	—	—
Luxembourg	—	—	—	—	—	—	—	—	—	—	—	—
Netherlands	—	170	144	190	167	297	428	597	657	748	1,006	1,724
Norway	—	—	—	—	—	—	—	—	—	—	—	—
Portugal	—	—	—	—	—	—	—	—	—	690	—	—
Spain	—	—	—	—	—	—	—	—	—	—	—	—
Sweden	—	—	—	—	—	—	—	—	—	—	—	—
Switzerland	—	—	—	—	—	—	—	—	—	1,425	—	—
Turkey	—	—	—	—	—	—	—	—	—	—	—	—
Yugoslavia	18	85	77	100	50	123	116	129	153	200	376	441
Total	685	6,637	8,298	9,027	9,817	12,748	15,818	18,289	19,959	22,985	23,029	25,407
United Kingdom	—	1	—	7	—	45	47	47	61	202	356	450
Eastern Europe												
Bulgaria	—	—	—	—	—	—	—	—	—	—	—	—
Czechoslovakia	21	210	356	995	1,079	1,903	1,848	1,860	1,526	1,425	1,265	987
E. Germany	—	—	—	—	—	—	—	—	25	30	40	50
Hungary	260	374	360	315	298	494	497	486	514	884	1,107	810
Poland	242	673	500	710	425	790	880	1,271	1,420	1,569	1,800	1,837
Rumania	4,023	6,870	8,402	10,010	9,319	12,604	14,138	15,000	17,626	18,776	20,500	23,016
Total	4,546	8,127	9,618	12,030	11,121	15,791	17,363	18,617	21,111	22,684	24,712	26,700
U.S.S.R.	7,300	10,656	15,200	25,703	33,900	48,677	61,840	79,249	95,850	118,443	143,267	—

*Unrevised figures.

Source: United Nations Statistics.

Internal consumption of commercial sources of primary energy in Europe, 1950 and 1955–65 (in '000 tons of coal equivalent)

Total Energy

	1950	1955	1956*	1957	1958*	1959	1960	1961	1962	1963	1964	1965
Western Europe												
Austria	10,885	14,210	15,993	16,362	16,040	15,958	17,860	17,814	19,366	21,715	22,794	24,344
Belgium	28,337	33,276	35,544	34,909	31,258	32,790	33,846	34,604	38,065	40,626	41,390	44,006
Denmark	8,896	11,208	11,337	9,940	10,067	10,970	12,700	13,356	15,147	16,846	17,823	19,952
Finland	4,148	6,596	6,838	7,930	8,031	7,671	8,723	9,918	11,258	11,004	12,309	13,311
France	83,458	101,917	114,192	114,684	115,536	112,519	120,349	124,968	133,584	145,327	151,771	162,977
W. Germany	125,064	171,800	186,060	189,618	185,206	177,930	204,271	209,776	236,688	253,773	262,985	262,133
Saar	6,782	8,861	9,522	9,597	9,501	9,381						
Greece	1,805	2,497	2,895	2,971	3,025	3,412	4,323	4,765	5,163	5,642	6,625	7,466
Italy	26,084	43,185	48,033	49,689	52,004	55,751	64,963	69,119	77,033	85,550	91,376	103,908
Ireland	3,239	3,554	3,741	3,111	3,376	3,614	3,712	3,978	4,076	4,294	4,565	5,050
Luxembourg	3,070	4,109	4,208	4,128	3,838	3,972	4,392	4,446	4,478	4,684	5,169	5,235
Netherlands	19,888	23,467	27,269	26,255	26,949	26,149	28,769	29,951	33,356	37,297	38,711	42,821
Norway	10,847	13,912	15,151	14,834	16,039	16,670	18,492	19,509	21,119	22,959	24,535	27,036
Portugal	2,132	2,827	3,323	3,305	3,759	3,664	3,924	4,367	4,613	5,955	5,285	5,746
Spain	16,224	20,954	22,619	25,520	26,988	27,993	29,173	29,801	32,262	34,864	37,335	41,255
Sweden	19,456	25,691	28,113	28,604	28,713	30,031	33,496	35,219	37,602	39,619	43,099	48,171
Switzerland	7,856	11,156	12,589	13,241	13,076	13,394	15,034	15,912	17,462	22,231	20,763	24,241
Turkey	3,845	6,026	6,347	7,040	7,588	6,849	7,293	7,256	8,382	9,728	10,608	11,071
Yugoslavia	8,166	11,303	12,850	13,957	14,397	16,101	17,814	18,569	19,657	21,868	23,415	25,179
Total	390,182	516,531	566,624	575,695	575,451	574,819	629,134	653,328	719,311	783,982	820,558	873,902
United Kingdom	223,662	245,996	253,093	244,203	243,594	233,768	251,270	253,781	258,918	270,389	272,269	289,275
Eastern Europe												
Bulgaria	3,314	5,684	6,189	7,377	7,832	9,502	11,008	11,645	14,270	15,586	17,299	18,000
Czechoslovakia	32,488	44,679	51,435	51,893	58,467	56,964	60,563	65,099	69,289	72,726	74,839	75,169
E. Germany	49,459	70,504	71,257	73,312	74,808	76,534	79,587	83,291	88,633	92,528	92,870	91,640
Hungary	9,077	14,777	13,595	15,742	16,004	17,191	18,383	19,843	20,273	22,997	25,362	24,455
Poland	51,581	71,535	78,097	83,022	81,942	87,754	92,404	96,076	100,088	106,283	110,520	111,465
Rumania	9,819	15,348	17,514	21,417	19,637	23,140	25,067	26,486	33,736	35,670	38,257	41,936
Total	155,738	222,527	238,087	252,763	258,690	271,085	287,012	302,440	326,289	345,790	359,147	362,665
U.S.S.R.	272,300	446,342	471,694	557,942	576,492	634,443	668,862	—	745,791	800,783	850,782	—

* Unrevised figures.

Source: United Nations Statistics.

SECTION III

SUPPLY AND DEMAND OF ENERGY IN EUROPE FROM THE PRESENT DAY UP TO 1980

I. Demand

Forecasts of the future energy demand, at least in the comparatively short or medium terms, are probably easier to establish and of a less speculative nature than forecasts with regard to the origin or sources of energy supplies. This is mainly due to the fact that Governments today are able to exercise a considerable degree of control over the growth rates in industry and of the economy as a whole and these are, in turn, determining factors in the rate of increase of demand for energy. Supply and demand forecasts covering all member countries of the former O.E.E.C. area for the period up to 1975 were made by both the Hartley and Robinson Committees, while in 1966 the O.E.C.D. Energy Committee made some tentative forecasts for as far ahead as 1980; by far the most detailed study of future trends in demand that has so far been attempted, however, was provided in 1963 by the three European Executives (the European Economic and Euratom Commissions and the High Authority) in their examination of the long-term energy prospects of the European Community,[1] in the elaboration of which the technical departments of the High Authority and the two Brussels Commissions worked in close co-operation with the interested government departments of the six member countries. More recently, the European Communities in their turn have sought to look a further five years ahead and carried their original estimates forward to 1980.[2] This report, however, deals only with the six countries that are members of the Common Market, and detailed forecasts of a similar nature do not exist for other European countries. In the first part of this section we have accordingly examined separately the demand estimates by sector of the Robinson and O.E.C.D. Energy Committee Reports, the exhaustive study made by the High Authority in conjunction with the

[1] *Etude sur les perspectives énergétiques à long terme de la Communauté européenne* Luxembourg, 1964.

[2] *Nouvelles reflexions sur les perspectives énergétiques de la Communauté européenne*, Luxembourg, 1966.

European and Euratom Commissions with regard to energy demand in the six Common Market countries, such information as is available for Eastern Europe, and probable trends in demand in the United Kingdom.

Sector forecasts in the Robinson and O.E.C.D. Energy Committee Reports

The estimates prepared by the Robinson Commission were based on a number of assumptions with regard to increases in the total labour force, the number of hours to be worked, productivity per man-hour, overall industrial production and Gross National Product, *i.e.*:

	1965	1975
Total labour force	104	106
Percentage of labour force employed	101	101
Working hours	97	95
Productivity per man-hour	133	179
Overall industrial production	147	208
Gross National Product	136	183

(1955: 100)

Thus, despite the very modest increase in total additional manpower availability, the Committee anticipated a rise of some 3% per year in productivity per man-hour—to be achieved as a result of the extensive application of automation and remote-controlled techniques—and an overall increase between 1955 and 1975 of 79%. The 1966 Energy Committee report based its estimates for 1970 and 1980, as we have seen, essentially upon an annual rate of increase in Gross National Production (henceforth G.N.P.) in the European area from 1964–1980 of between 4 and 4·5%.

The iron and steel industry

The Commission's estimates of domestic and steel consumption were projected from the index of overall industrial production, by elasticity coefficients of 1·25 for 1955–65 and 1·15 for 1965–75. Exports were expected to rise during the first ten years up to 1965 but to fall back subsequently to about their 1955 level. 'Requirements of fuel (other than coke for pig-iron production) were projected to increase in line with domestic steel production. Coke requirements were estimated separately, using the projections of pig-iron production. An allowance was made for some decline in the coke-pig-iron ratio in blast furnace operation due to advances in production techniques.'[1] The figures retained by the Commission for steel production and energy demand were as follows:

	1965	1975
Steel production (1955: 100)	155	217
Energy requirements (in 10^{12} kcal)	728	998

This compared with total energy requirements by the industry in 1955 of 484×10^{12} kcal.

General industry

The rate of increase of industrial production in this sector was estimated by excluding the projected growth of the iron and steel industry from the index of overall industrial production. It was then assumed that the energy requirements of the sector would rise at about the same rate as its actual output. 'It was felt that the general trend to higher mechanisation of the production apparatus would continue, and that a particularly rapid growth was likely to be registered by certain energy intensive industries (electro-chemistry, metallurgy)'.[2] The Commission's estimates were accordingly as follows:

	1965	1975
Industrial production (excluding iron and steel) (1955: 100)	146	206
Energy requirements (in 10^{12} kcal)	1,366	1,754

This compared with total energy requirements in this sector in 1955 of $1{,}049 \times 10^{12}$ kcal.

Transport

This sector was subdivided into road transport and aviation, rail transport and water transport. The Commission drew particular attention to the changing pattern of demand in the rail transport section, where the change-over from steam to electric and diesel locomotives was at that time still gathering momentum in many of the member countries. As a result it was estimated that although useful energy requirements would increase by some 30% between 1955 and 1975, the gross demand for energy would in fact decline by two-fifths. The estimates with regard to road transport were based on projections of the stock of vehicles and on expected trends in utilisation of private and commercial vehicles. For aviation, the Commission estimated that requirements of energy would treble by 1965 and then double again in the succeeding years. Estimates for water-traffic in inland waterways and canals were based on G.N.P. growth rates and bunker requirements for ocean-going vessels on the anticipated rate of development in trade between Western Europe and the rest of the world. The Commission's total estimates for the sector were accordingly as follows:

[1] *Towards a New Energy Pattern in Europe*, *op. cit.* p. 108. [2] *Ibid.* p. 108–9.

Useful energy demand (1955: 100):		1965	1975
Road transport and aviation		168	251
Rail transport		115	131
Inland waterways		134	186
Ocean transport		139	195
Energy requirements for (in 10^{12} kcal)	**1955**	**1965**	**1975**
Road transport and aviation	288	495	725
Rail transport	272	229	157
Inland waterways	52	48	51
Ocean transport	149	197	270
	761	969	1,203

Domestic sector

This sector included energy consumption by private households (accounting for between 60 and 65 % of total consumption), by small industries and crafts (about 15 %), and by agriculture, local authorities, government and commercial premises. 'The projection for household fuel consumption was derived from the estimates of private consumption expenditure on fuel and light. Within this subsector, the application of approximate relative prices for individual fuels to the consumption pattern projected for the future resulted in the conclusion that the forecast increase of 80 % in expenditure by 1975 would correspond to an increase only one quarter to one third as high in the total quantity of gross energy consumed.[1] In the small industries sector, gross energy demand was assumed to increase slightly less quickly than energy demands in the "other industries" sector. For the remaining subsector, it was assumed that an increase of 0·7–0·8 % in the energy demand of the agricultural, commercial and other consumers would be associated with each increase of one per cent. in the index of value added in the agricultural and services industries.'[2] The retained estimates for the sector were as follows:

	1965	1975
Household fuel and light, expenditure (1955: 100)	136	182
Household fuel and light gross energy demand (1955: 100)	112	127
Small industries gross energy demand (1955: 100)	125	156
Agriculture, commerce, government, local authorities, etc. gross energy demand (1955: 100)	123	151
Total energy requirements of the sector (in 10^{12} kcal)	1,691	1,960

This compared with total energy requirements of $1{,}470 \times 10^{12}$ kcal in 1955.

[1] This is due to the fact that electricity consumption was projected to increase very rapidly. Since electricity is relatively much more expensive than other fuels per unit of energy the expenditure on energy increases more rapidly than the quantity of gross energy consumption.

[2] *Towards a New Energy Pattern in Europe*, *op. cit.* p. 110.

Conclusion

The aggregate of the four sector estimates (i.e. iron and steel industry, general industry, transport and the domestic sector) resulted in an estimated total demand for energy in 1965 and 1975 of 4,754 and 5,915 × 10^{12} kcal respectively:

Table 41. *Energy demand by end-use sectors in 1955, 1965 and 1975 (in 10^{12} kcal)*

	1955	1965	1975
Iron and steel industry	484	728	998
General industry	1,049	1,366	1,754
Transport	761	969	1,203
Domestic and miscellaneous	1,470	1,691	1,960
Total	3,764	4,754	5,915

When the losses incurred by transformation and the energy industries' own consumption requirements were taken into account, the total requirements of primary energy expressed in terms of coal equivalent were estimated to reach 1,010 million tons in 1965 and 1,340 million tons by 1975.

With regard to demand for secondary energy, the Robinson Commission forecast an increase in consumption of electricity of almost 100 % between 1955 and 1965, from 358 to 710 TWh, and a further rise of 70 % between 1965 and 1975, from 710 to 1,200 TWh; consumption of gas was expected to rise from 65 billion m^3 in 1955 to 100 billion m^3 in 1965 and to between 160 and 260 billion m^3 by 1975 (the exceptionally wide range in the estimated consumption figure for 1975 was explained by the degree of uncertainty that then existed with regard to the extent of indigenous natural gas resources). In both cases, that is, electricity and gas, a considerable part of the very high rate of growth in consumption was due to substitution within existing or expanding uses for alternative forms of primary or secondary energy.

The 1966 O.E.C.D. Energy Committee Report was far less concerned than its predecessor with detailed study of the trends in energy requirements in individual consumer sectors. While showing figures for consumption in major sectors for 1970 and 1980 the figures were not broken down or disclosed in any detail.

The 1966 Report emphasized once again the uncertainty and unreliability of long-term energy forecasting. Just as the Robinson Report served to show the extent to which the Hartley Commission had completely misjudged the forward position with regard to coal production and demand, so the 1966 Report revealed how widely the Robinson Report had miscalculated the forward total demand for energy. Whereas, for example, the Robinson Report has estimated a total energy demand in 1965 of 1·010 million tons of coal equivalent, actual consumption in 1964 had already reached 1·073 million tons, while the Energy Committee's forecast of demand in 1970 was 1·370 million tons—30 million tons more than its predecessor's estimate for 1975.

Table 42.* *O.E.C.D. Europe: Estimates of Energy Requirements according to Main Sectors of Demand*

	Million tons of coal equivalent		
	1964	1970	1980
Final consumption in Iron and Steel	94	108	126
Other industry	225	280	402
Transport (incl. bunkers)	170	230	348
Domestic and miscellaneous	276	342	510
Energy Sector	93	110	154
Total consumption	858	1,070	1,540
Transformation losses and non-energy products	215	300	510
Total primary energy requirements	1,073	1,370	2,050

* 1966 O.E.C.D. Energy Committee report, *op. cit.* Annex III, Table D.

The same remarks apply to all the sector forecasts in the Robinson Report. Broadly speaking, we may say that the consumption figures contained in the Robinson Report will have been reached two to four years sooner than had originally been anticipated. In the electricity sector, for instance (admittedly an exceptional case), the Robinson Report's forecast of a consumption figure of 710 TWh in 1965 had already been surpassed by nearly 50 TWh or some 7% in 1964 and was expected to reach 1·570 TWh in 1970 against the original estimate in the Robinson Report of 1200 TWh by 1975.

The position is very different with regard to the long-term energy forecasts prepared by the European communities. First published in 1964 these were the result of prolonged and intense research and discussion between the High Authority and the European and Euratom Commissions and officials in the responsible Governments in the six Community countries, and included analytical studies of the main energy markets.

Sector forecasts for the six Common Market countries

In making their detailed forecasts in 1963 of energy demand in the six Common Market countries up to 1975, the three European Executives relied heavily on the historical degree of correlation between increases in G.N.P. and industrial activity and rising demand for energy. As we have already seen in the case of the estimates prepared by the Robinson Report, future demand is not such an unknown factor or quantity as might at first sight appear. Not only are basic growth patterns unlikely to show dramatic or sudden changes, particularly in days when the primary objectives of most Governments are full employment and a rising standard of living by means of increasing G.N.P.; but the effect of major new discoveries in fuel techniques is cushioned by the inevitable delay between their development in the laboratories and their commercial application on a wide scale.

The fundamental statistics retained and used by the three executives and upon which all their estimates in the last resort were based were therefore those relating to the rate of increase in G.N.P. and the rate of increase in industrial production in the Common Market countries between 1960 and 1975.

Table 43. *Estimated rates of increase in G.N.P. and industrial activity in the Common Market countries: 1960–1975*

	1960–65	1965–70	1970–75
Rate of annual increase in G.N.P.			
Belgium	3·8	3·9	3·9
France	5·2	4·7	4·6
Germany	4·4	4·0	4·2
Italy	5·95	5·75	5·3
Netherlands	4·3	4·9	4·7
Rate of annual increase in industrial production			
Belgium	4·8	4·8	4·8
France	5·6	5·9	5·5
Germany	5·5	5·0	5·0
Italy	8·8	7·8	6·5
Luxembourg	4·0	4·0	4·0
Netherlands	5·4	6·0	5·6

The 1966 long-term study estimated that G.N.P. in the Community as a whole would rise at a rate of 4·6% a year between 1965 and 1970; the rate of increase was expected to be somewhat above the average in France, Holland and Italy; somewhat below in Belgium, Germany and Luxembourg. General industrial production was expected to show a slightly faster rate of increase. The study frankly admitted that there were up to that time only some rather fragmentary indications for the period 1970–1980 but suggested that, as a working hypothesis, the same figures used for the earlier period might be applied.

In addition to the correlation between increases in G.N.P. and industrial production and rising demand for energy, the energy experts of the three European Executives also made a careful examination of consumption forecasts by individual sector. They were, however, emphatic that the figures published in the *Perspectives Energétiques à Long Terme* should be regarded as indications of trends rather than as hard and fast forecasts. They also stressed the fact that there were certain elements of uncertainty, notably fluctuations in world trading activity which were to a large extent beyond the control of individual Governments, which could affect their estimates by 10–15% in a downward or up to 10% in an upward direction. Furthermore, their figures were based on average conditions of climate and hydraulicity—and exceptional conditions in either of these two factors could result in a variation of up to 5% from the mean. It is however only fair to comment that it is most unlikely that all three of these factors would exert effects tending in the same direction at any one time; a good deal of cancelling out could therefore be expected. With these reservations in mind we can now pass on to consideration of the individual consumption sectors, where we have deliberately followed the same sequence as in the case of the Robinson Report.

The iron and steel industry

Between 1950 and 1960 steel production in the Common Market countries increased at the exceptionally high rate of 8·7% per year—more than 3% above the annual growth rate in G.N.P. In the light of the continued high rate of growth both in general industrial activity and in G.N.P. that was anticipated throughout the 15-year period from 1960 to 1975, the energy experts of the European Executives estimated that steel output would continue to show a substantial rate of increase, albeit at a somewhat slower tempo than in the 1950–60 decade. Thus, for the period 1960–70, a rate of growth of 4·2% was forecast, that is, somewhat below the increase in G.N.P. which had been estimated at 4·76%. Estimated steel production for the six Common Market countries for the period 1960–75 is shown in Table 44.

By 1966, however, it had become evident that these estimates were considerably over optimistic and that the actual level of steel production in the six Common Market countries was unlikely to exceed 95 million tons in 1970. No figure was indicated for 1980 but, given the present surplus all over the world in steel producing capacity, it may well be that the actual 1980 production capacity in the Community will not exceed 110 to 115 million tons.

Table 44. *Steel production in the six Common Market countries: 1960–75 (in '000 metric tons)*

	Actual		Estimated	
	1960	1965	1970	1975
Belgium	7,181	9,200	10,100	11,000
France	17,299	19,600	26,900	31,500
Germany	34,101	36,800	47,500	57,500
Italy	8,229	12,700	16,800	20,000
Luxembourg	4,083	4,600	4,900	5,200
Netherlands	1,942	3,100	3,800	4,800

Specific coke requirements (*i.e.* per ton of steel produced) were expected to fall substantially during this period. This assumption was based on the fact that, although specific consumption of coke in blast furnaces had remained relatively stable throughout the period from 1950 to 1957, there had subsequently been a steady and continuing improvement. Thus, by 1960, average specific coke consumption per ton of steel produced in the six countries had fallen from 972 to 883 kg. and this was expected to fall further to 750 kg. in 1965, 670 kg. in 1970 and 640 kg. by 1975. This was equivalent to a 15% reduction in specific coke consumption between 1960 and 1965, 10% in 1965–70 and 5% in 1970–75. In fact, the actual figure in 1965 was 700 kg. and the 1966 estimates indicated an average specific coke consumption of 600 kg. in 1970 and 500 kg. by 1980. Thus further progress in this field was expected to continue steadily and progressively throughout the period under review. One factor, which may ultimately decisively affect coke consumption by the steel industry, is the preliminary experimental work now being carried out in Germany and elsewhere on the use of atomic energy in blast furnaces: this would enable blast furnaces to use coking fines directly and so eliminate the need for coke entirely. It is unlikely, however, that this will be a major factor during the period under review.

If the other non-electrical energy requirements of steelworks and rolling-mills are included, then the total energy requirements of the steel industries of the Common Market countries may be expected to rise from their 1965 level of 61 million tons of coal equivalent to 66 million tons by 1970 and 74 million tons by 1980.

General industry

While consumption of non-electrical energy by the general industry sector in the Common Market countries increased during the period from 1950 to 1960 at an average annual rate of 4·2 %, there were, inevitably, considerable variations from this rate in individual countries: thus, in Italy, consumption of non-electrical energy in this sector rose at an annual average rate of no less than 10·8 %; next came Luxembourg and the Netherlands with 4½ %, Germany with 4 %, France with 3·3 % and, bringing up the rear, Belgium where there was actually a fall of 2·6 %. These remarkable differences are explained by the varying rates of increase in the use of electricity and in its substitution for other fuels, while, at the same time, comparisons of trends in the six countries are complicated by the very different structures, or breakdown of the energy components, of the general industry sector in the six countries. These differences were expected to continue into the period 1960–75 but at a less pronounced and gradually reducing rate. Thus, in Germany, where the level of general industrial output was expected to increase between 1960 and 1975 at a lower rate than between 1950 and 1960, energy requirements in this sector were not expected to increase at a rate exceeding 2·3 % per year. In Belgium, general industrial expansion between 1960 and 1975 was expected to increase at the same high rate as in the boom years of 1950–55, leading to a total demand in this sector of 5·8 million tons in 1970 and 6·5 million tons by 1975. In France, the general level of industrial activity was expected to continue at about the same rate as in the period 1950–60, resulting in a total demand of some 32 million tons in 1970 and 36·4 million tons by 1975. In Italy, where as we saw, consumption of energy showed a dramatic rise in the 1950–60 decade, energy demand was expected to rise about 6·8 % between 1960 and 1970 and a slightly lower rate between 1970 and 1975. Steady increases in energy consumption were also expected in the Netherlands and in Luxembourg, although in both cases the rate of increase was likely to be well below the corresponding G.N.P. figures.

As far as individual sources of energy are concerned, oil was generally expected to have established a dominating position in this sector by 1970. Thus, it was expected that by 1970 oil would hold 75 % of the market in the Netherlands, 70 % in Belgium, 65 % in Luxembourg, 55 % in Italy and Germany, and 52 % in France. The share of solid fuel was estimated at no more than 27 % in Germany (compared with 63 % in 1960), 20 % in France, 15 % in Luxembourg, 9 % in Belgium, 5 % in the Netherlands and nil in Italy. Gas was forecast as providing about 40 % of the energy requirements in this sector in Italy and about 20 % in the other five countries. Total non-electrical energy requirements in the Common Market countries were estimated at 125 million tons in 1970 and 143 million tons in 1975.

While the Community study admitted that the final share of the total energy requirements of the general industrial sector that would fall to each fuel was subject to a number of factors, any one of which could substantially affect the balance set out in the preceding paragraph, it was firmly believed that there was a ceiling to the demand for coal; the level of this upper limit would be determined by the requirements of the metals, textiles and chemical industries[1] and might, under exceptionally favourable circumstances, enable coal to retain some 45 % of this market in the Common Market countries as a whole in 1970 and about 40 % in 1975—in this event, coal's share of the market would be about 50 % in France, Germany and Luxembourg, about 30 % in Belgium and the Netherlands and 15 % in Italy. The Community experts, however, were more inclined to believe that the swing from coal to oil that has been so pronounced in this sector in recent years would continue, particularly if—as they thought likely—enterprises decided to replace obsolescent coal-burning equipment by oil or gas-fired apparatus. They accordingly estimated the share of coal and coke by 1970–75 at between 50 and 55 million tons (with the lower figure as the more probable), that of lignite at 4 to 5 million tons of coal equivalent, that of gas—including both manufactured and natural gas—at between 25 and 40 million tons and, finally, that of oil at between 65 and 100 million tons of coal equivalent (with the higher figure in this case being the more probable).

The 1966 study concluded that the rate of improvement in fuel utilisation, which had been such a feature during the period 1950–1960, was likely to slow down substantially during the coming decade. At the same time the rate of substitution of oil and natural gas for coal was expected to continue at a rapid rate, so much so that consumption of solid fuels in the general industrial sector was now expected to fall to between 15 and 22 million tons in 1970 and under 10 million tons by 1980. Since the overall energy requirements in this sector were not expected to show any appreciable change from the original 1963 estimates, any decline in consumption of solid fuels would inevitably lead to an increased demand for oil or natural gas.

[1] It being assumed that the petro-chemical industry will by 1970 have gone over completely to oil.

Transport

This sector was sub-divided into road traffic, railways, inland navigation and air traffic. (The Community study differed in this respect from the O.E.C.D. Energy Advisory Committee which, it will be recalled, included ocean bunkers in the transport sector: the Community experts contended, rightly, that this fell outside the scope of a study on the internal energy demand of the Common Market countries).

The number of both private and commercial vehicles in the Common Market countries was expected to continue to increase at a rapid rate throughout the period from 1960 to 1975; thus, the number of private vehicles (including small vans) was expected to rise from 12¾ million units in 1960 to over 47 million by 1975 (the biggest relative increase was expected to take place in Italy: from 1¾ million units in 1960 to 11 million by 1975; while the biggest absolute increases were expected to occur in Germany—4½ million to 16 million—and France—5¼ million to 15 million). At the same time the fleet of commercial vehicles (*i.e.* lorries, road-trucks, road-tankers and articulated lorries) was expected to double in size from 3¼ million units in 1960 to 6¾ million by 1975 (the main contributor in this case being France where the number of commercial vehicles was expected to reach 3¼ million units—virtually half the Common Market total—by 1975). As a result, consumption of petrol and gas-oil was expected to increase between 1960 and 1975 from 15½ to 42¾ million tons and 7¼ to 20½ million tons respectively.

The pattern of the non-electrical energy requirements of the railways will be determined by the rate at which the use of coal and oil is eliminated. The present trend in all six countries is towards the electrification of high-density traffic passenger and freight lines, dieselisation where electrification is not economically viable, closure of marginal lines and, in certain cases, substitution of road for rail traffic. At the time of writing however the overall pattern in the six countries still shows very wide differences; thus, in Germany, coal is still an important source of fuel for the railways (with coal requirements in 1965 estimated at 4·7 million tons against only 350,000 tons of diesel oil and 300,000 tons of fuel oil), mainly for short-distance freight hauls; a big electrification programme is in hand but it is not expected that this will be completed before 1970, while coal may continue to be used, although in steadily diminishing quantities, until 1975. In Belgium steam traction is expected to have been completely replaced by electric and diesel traction by 1967/68. In France, electric traction already accounted for some 57 % of all rail traffic in 1960 and this percentage will rise to 70 % by 1968; steam traction is expected to continue for short freight hauls until 1969/70. In Italy, the optimum level of electrification for both passenger and freight traffic has already been attained and the few remaining steam engines will shortly be replaced by diesel engines. In Luxembourg diesel traction predominates, although the main lines to Belgium, France and Germany are in process of being electrified; a small amount of steam traction may remain until 1970. Finally, in the Netherlands, the steam engine was taken out of service in 1958: today 50 % of the lines are electrified while the remainder are served by diesel traction.

Inland waterways constitute an extremely important form of transport throughout the whole of northern Europe and, in particular, on both sides of the Rhine, in Holland, Belgium and northern France—where there is an extensive and important network of canals—and northern, and parts of central, Germany. Both in Germany and in Holland coal is widely used as a form of fuel and is not expected to have been completely displaced by other fuels before 1970. In France and Italy fuel oil is widely used. The general tendency, however, is to shift over to diesel-oil and this is expected to account for about 90 % of the fuel requirements of this sector by 1970.

Turning to air traffic, the Community experts considered that this sector had in recent years been dominated by three main factors: an exceptionally rapid increase in both passenger and freight traffic, accelerated technical progress with regard to commercial aircraft, and a multiplication of air transport companies. All three factors had combined to cause a substantial increase in the number of aircraft—dramatically illustrated by the build-up of the big jet fleets, often for reasons of national prestige. As a result, fuel requirements in this sector in the Common Market countries were expected to rise from 1·4 million tons in 1960 to some 5 million tons in 1970 and as much as 10 million tons by 1975.

As a result total fuel requirements in the overall transport sector in the six Common Market countries were expected to increase from some 30 million tons in 1960 to 109 million tons in 1970 and 164 million tons by 1980.

Domestic sector

As in the case of the O.E.C.D. study, this sector embraced households, local and Government services, handicrafts and agriculture. Annual consumption of non-electrical energy during the period 1950 to 1960 had risen by widely varying amounts in the six Common Market countries, that is, by 9·7 % in Italy, 5·3 % in Germany, 3·2 % in Luxembourg, 3 % in France, 2·3 % in the Netherlands and 0·1 % in Belgium. Expressed in terms of consumption per head of population, demand throughout this period was highest in Belgium and Luxembourg (with 1,027 and 1·147 kg per person respectively in 1960) with Germany, the Netherlands and France in the next three places. Italy, with a *per capita* consumption figure of 73 kg in 1950 and 184 kgm in 1960 came a long way behind in sixth and last place. The increase in consumption of energy in the domestic sector in the six countries as a whole from 1950 to 1960 had been substantially below the G.N.P. growth rate (*i.e.* elasticity of 0·72) and this was expected to fall to a considerably lower level still during the period from 1960 to 1975, that is, 0·55 from 1960 to 1965, 0·40 from 1965 to 1970 and 0·20 from 1970 to 1975. Here again, however, in Belgium and Germany, where general living standards by 1960 were already high, little increase in energy demand was expected. Similarly, in the Netherlands, the increase in demand was barely expected to exceed 1 % per year. In France and Italy on the other hand a much higher rate of increase was anticipated, leading to elasticities (in relation to the G.N.P. growth rate) of 0·65 (1960–65), 0·60 (1965–70) and 0·55 (1970–75) in the case of France and 1·35, 1·15 and 1 in the case of Italy. As a result total demand for non-electrical energy in this sector in the six countries was expected to rise from a total of 96½ million tons of coal equivalent in 1960 to 150½ million tons by 1975, with comparable *per capita* consumption figures of 572 and 704 kg.

It was expected that saturation point would be reached in this sector in all six countries during the period under review. This meant that a substantial part of such increases in demand as were estimated (except, as we have already seen, in the case of Italy and France) could only be accounted for by population increases. In Italy the comparatively low standards of living in the south of the country and in Sicily had previously resulted in the burning of large quantities of wood, but this variation from the general Community pattern was expected to disappear with rising standards of living and the growing use of more sophisticated forms of energy. Differences in climate had also to be taken into account, although the effect of such variations in the six Common Market countries was stated to be of less importance than was often popularly supposed.

The 1966 report, however, found that, contrary to the expectation expressed in the earlier study, consumption of fuel in the domestic sector was rising rapidly and that there was no evidence that the market was anywhere near saturation. Energy consumption in this sector was consequently expected to rise from the actual level of 139 million tons in 1965 to 165 million tons in 1970 and 220 million tons by 1980. The reason given for this previously unsuspected market development was that, with growing prosperity throughout the Community, people were both able and prepared to spend a great deal more on their homes. Thus, throughout the northern part of the Common market area people were no longer content to live in one or two rooms during the long cold winter months, but wished to use the whole of their house. Moreover, they were prepared to spend heavily for this greater level of comfort and convenience.

The shares of the domestic sector market held by the various fuels were expected to show substantial changes over the period under review. Thus consumption of solid fuels in the Community area as a whole was expected to fall from a total of 64¾ million tons (some 67% of the total energy requirements of the domestic sector) in 1960[1] to 60 million tons in 1965 and 40 million tons by 1975—although extreme upper limits of 45–50 million and 75–80 million tons were conceded for consumption of coal, coke and lignite in these same two reference years under exceptionally favourable circumstances for solid fuels. At the same time demand for gas and oil was expected to increase to levels of between 15 and 20 million tons (*i.e.* 12–13% of the market) and 35–70 million tons (with the higher figure being regarded as the more probable) respectively by 1975.

The 1966 study forecast a level of consumption for solid fuels of 35 to 45 million tons in 1970 and substantially less by 1980. Consumption of oil and natural gas on the other hand was expected to rise steeply, with natural gas covering some 40% of total fuel requirements in this sector by 1980.

Power-generating

Between 1950 and 1960 the consumption of electricity in the six Common Market countries grew by an average of 9·4% a year—equal to a doubling of demand every seven years. As power-generating constitutes the most rapidly growing sector in demand for energy—owing to the seemingly insatiable appetite that exists not only in the Community but which has manifested itself in all other advanced economies—we have followed the pattern of the original Community study by taking each main consumption sector in turn to examine the likely rate of growth in demand for electric power:

(i) *Iron and steel industry:* Consumption of electricity in the Community area in this sector in 1960 amounted to 27·4 TWh, compared with

[1] This was despite the fact that consumption of solid fuel fell much less rapidly during the period 1950–60 in the domestic than the general industrial sector, due to a general reluctance in Belgium, Germany and northern France to part with the traditional coal stove and a good deal of inertia.

10·4 TWh in 1950, that is, a rate of increase of some 10·2 % per year. Consumption of electricity per ton of steel produced rose from 327 kWh in 1950 to 376 kWh in 1960 (this figure disguised substantial differences from one country to another, *i.e.*, Belgium with a consumption figure of about 320 kWh per ton of steel remained fairly stable throughout this period, but in Italy and the Netherlands there were in fact substantial decreases—due essentially to improvements in energy utilisation—from 762 to 632 kWh and from 408 to 309 kWh respectively).

The estimated level of steel output for the six Common Market countries of 89 million tons by 1965 was expected to result in an increase in electricity requirements of 8·9 TWh, leading to a total demand by that year of 36·3 TWh and a rate of electricity consumption per ton of steel produced of 408 kWh. After 1965 and up to 1970 demand for electricity was expected to increase at a rate of about 1 % for every additional ton of steel produced: this rate was expected to fall to an increase of 0·6 % for every additional ton of steel in the next five year period from 1970 to 1975; leading to total electricity requirements by the steel industry of 47·2 and 57·2 TWh respectively and a rate of electricity consumption per ton of steel produced of 430 and 443 kWh respectively in the same two years. Generally speaking, demand was expected to rise annually by about 2 % in Belgium, France, Germany and Luxembourg, by 1 % in the Netherlands and to fall further by about 1·5 % in Italy.

(ii) *General industry:* Consumption of electricity by general industry in the six Common Market countries in 1960 amounted to 129 TWh, that is, about 45 % of total consumption of electricity in that year. It is worth noting that this proportion was roughly similar in all the member countries with the single exception of Luxembourg. The average rate of increase in consumption between 1950 and 1960 was about 8·5 %, that is, slightly higher than the rate of increase in industrial production in this sector. This rate of increase was expected to continue throughout the period under review, albeit with minor variations from one country to another, *i.e.*, 8·7 % in Italy, 8·3 % in the Netherlands, 8·1 % in France and 7 % in Belgium, Germany and Luxembourg. As a result total demand for electricity in the general industry sector in the Community area was expected to rise from a level of 129 TWh in 1960 to 381·5 TWh by 1975.

(iii) *Transport:* Consumption of electricity by the transport sector in the six Common Market countries in 1960 amounted to 11·8 TWh, of which electrified railway lines accounted for 8·5 TWh and local transport, *i.e.*, trams, for the remainder. The transport sector thus accounted for only 4·1 % of the total consumption of electricity in the Community area, although the precise figure varied between 3 and 6% according to the individual country in question. Consumption of electricity in this sector increased between 1950 and 1960 at an annual average rate of 6·8%, *i.e.*, below the rate of increase obtained in most other sectors, due mainly to keen competition from other fuels, notably diesel oil. It must be borne in mind that electrification of railway lines is economically viable only in cases where there is high density passenger, and to a lesser degree also freight, traffic. Furthermore, by 1960 the electrification process had already been virtually completed in Italy and the Netherlands and was well advanced in France.

The Community experts estimated that railway traffic would rise at a much slower rate in the period from 1960 to 1975 than either G.N.P. or industrial output, due mainly to consumer and producer preferences for alternative forms of transport. As a result rates of increase in the demand for electricity by the transport sector for the Community area were estimated at a maximum of 5½% and probably less, varying from 7·2% in France and 5·3% in Germany to little over 3% in Italy and the Netherlands.

(iv) *Domestic sector:* Consumption of electricity in the domestic sector in the six Common Market countries in 1960 amounted to 59·2 TWh, or about 21% of overall electricity consumption. Once again this global Community figure disguised substantial variations from one country to another, *i.e.*, 28% in the Netherlands, 16% in Belgium and 12% in Luxembourg. Within the domestic sector grouping, households accounted for about 55% of the total. *Per capita* consumption of electricity in this sector in the Community area, which in 1960 at 350 kWh was only about a third of the United Kingdom level, varied from 618 kWh in Luxembourg and 484 kWh in Germany to 260 kWh in Belgium and 239 kWh in Italy; this was mainly due to the comparatively limited use of electrical household goods and apparatus (i.e. vacuum-cleaners, electric cookers, television sets, etc.) in the Community area. Despite this very big disparity, consumption of electricity in this sector rose between 1950 and 1960 by some 10% per year and was expected to continue to rise at a rate not much below this figure during the whole of the period under review although at a less rapid rate in Germany than in the five other countries. As a result total consumption of electricity in the domestic sector in the Community area was expected to rise from 59·2 TWh in 1960 to 90·7 TWh in 1965 and 198·2 TWh by 1975.

(v) *Energy transformers and distributors:* Consumption of electricity in this sector in the six Common Market countries in 1960 amounted to 57·4 TWh, or 20% of total consumption of electricity. The Community experts estimated that the losses incurred in transporting and distributing electric current would be considerably reduced in the period under review as a result of the trend towards building large new power stations near to the major centres of demand; this process would be further encouraged by the development of high tension transmission lines which, while still costly, made a major contribution towards reducing losses of current in this sector from 125 Wh/kWh in 1950 to 89 Wh/kWh in 1960—an annual rate of improvement of some 3·5 %. The maintenance of a similar rate of improvement throughout the period 1960 to 1975—which was considered to be technically feasible—would have the effect of further reducing losses to a level of 65 Wh/kWh (equivalent to a total of some 37 TWh).

The addition of the results obtained in the examination of future electricity requirements of the major consumption sectors gave a total demand figure for the Community area as a whole of 284·8 TWh in 1960; this was expected to rise to 408·9 TWh in 1965, 573·9 TWh in 1970 and 788·6 TWh by 1975. Within this global figure, general industry was expected to account for some 48 % of total requirements and the domestic sector for a further 25 %. On a country basis the leading consuming nation was Germany, with 316·6 TWh or 40 % of total, followed by France with 217·6 TWh, or 28 %, and Italy with 165·8 TWh, or 20 %:

Table 45. *Estimated total demand for electricity in the six Common Market countries by country and by sectors: 1960–75 (in TWh)*

	Actual		Estimated	
	1960	1965	1970	1975
A. By sectors				
Iron and steel industry	27·4	38·5	47·2	57·6
General industry	129·0	174·0	272·2	381·5
Transport	11·8	15·7	19·7	23·8
Domestic sector	59·2	108·2	135·3	198·2
'Energy' sector	57·4	89·5	99·5	127·5
Total	284·8	425·9	573·9	788·6
B. By countries				
Belgium	15·2	22·2	27·0	35·7
France	74·8	109·8	154·6	217·6
Germany	120·6	181·4	234·4	316·1
Italy	56·1	83·4	119·3	165·8
Luxembourg	1·57	3·9	4·37	4·90
Netherlands	16·5	25·2	34·2	48·5
Total	284·8	425·9	573·9	788·6

In their 1966 study the Community experts forecast a continued rapid rate of increase in demand for electricity, which was expected to double once again during the course of the next ten years. Thus, electricity requirements were expected to increase between 1965 and 1980 by 280% in the general industrial sector, by 400% in the domestic sector, and by 290% in total. These forecasts, if realised, would mean an increase in demand for electricity in the Common Market countries from 426 TWh in 1965 to 1214 TWh by 1980. Such an expansion in demand would far exceed the estimated increase in the years after 1970 in the original 1963 Community estimates and would require an enormous investment programme in new power station construction. In terms of tons of coal equivalent, energy consumption by the power station sector could be expected to increase from 139 million tons in 1965 to 165 million tons in 1970 and 386 million tons by 1980.

The fuel requirements of the power-generating industry in the Community area

In their forecasts with regard to the fuel requirements of the electricity-generating industries in the six Common Market countries, the Community experts in their 1963 study had assumed as working hypotheses that the economically viable hydro-electric resources of Germany and Italy had already been largely exploited, that some 50–60% of total French resources had already been harnessed and that electric current from nuclear power stations would not become a significant factor until after 1970 (and, even then, only in comparatively small quantities). As a result the total amount of electricity to be produced during the period under review from each of the main groups of power plants, that is, hydro and geothermal, nuclear and conventional thermal stations, was expected to develop as shown in Table 46.

Thus, throughout the period under review, conventional thermal plants were expected to have to bear the brunt of production of electricity in the Community area, despite steadily increasing contributions—in absolute if not in relative terms—from hydro-electric and nuclear plants. At the same time the size of power stations was expected to increase, from units of 50–300 MW as at present to giant plants of between 1,000 and 1,500 MW; a development which it was generally believed would lead to a fall in the specific consumption of fuel per unit of electricity produced of between 4 and 5%. It is worth noting that the hydro-electric power stations in France and Italy are generally called upon to provide base-load requirements—in the other four countries base-load supplies are necessarily

Table 46.* *Estimated production of electricity in the six Common Market countries from hydro, nuclear and thermal power stations: 1960–75 (in TWh)*

	1960 (actual)		1965 (estimated)		1970 (estimated)		1975 (estimated)	
	Total	%	Total	%	Total	%	Total	%
			(a) Hydro and geothermal plants					
Belgium	0·2	1·3	0·3	1·5	0·3	1·1	0·3	0·8
France	40·9	54·9	43·5	40·3	51·0	33·0	58·0	26·6
Germany	13·0	11·1	15·0	9·0	19·5	8·5	21·0	6·8
Italy	48·2	85·9	51·0	61·2	57·5	48·2	63·0	38·0
Luxembourg	—	—	1·1	37·9	1·4	40·0	1·4	40·0
Netherlands	—	—	—	—	—	—	—	—
			(b) Nuclear plants					
Belgium	—	—	0·1	0·5	0·5	1·9	2·0	5·6
France	0·2	0·2	2·5	1·9	9·2	5·9	25·0	11·5
Germany	—	—	0·2	0·2	4·0	1·7	25·0	8·0
Italy	—	—	3·7	3·6	6·0	5·0	2·6	15·7
Luxembourg	—	—	—	—	—	—	—	—
Netherlands	—	—	—	—	0·3	0·9	2·0	4·1
			(c) Thermal plants					
Belgium	15·0	98·7	19·6	98·0	26·0	97·0	33·0	93·6
France	34·0	44·9	62·0	57·8	94·0	61·1	135·0	61·9
Germany	103·4	88·9	150·8	90·8	206·0	89·8	265·0	85·2
Italy	8·0	14·1	28·3	35·2	56·0	46·8	78·0	46·3
Luxembourg	1·5	100·0	1·8	62·1	2·1	60·0	2·1	60·0
Netherlands	16·5	100·0	24·0	100·0	34·0	99·1	46·0	95·5

* This table does not take into account exchanges of current between countries, *i.e.* between Germany and Austria and Switzerland or between Germany and Scandinavia—the Konti-Skan project.

generated by normal conventional plants. Despite certain advantages in the case of hydro-electric plants, the vast bulk of new power stations coming into operation in the period 1960–75 as well as the stations providing the lion's share of overall electricity requirements, were expected to be conventional thermal plants. As a result fuel requirements (*i.e.* requirements of the conventional thermal power stations) in the Community area were expected to increase from 80·6 million tons of coal equivalent in 1960 to 104·2 million tons by 1965 and 185·5 million tons by 1975.

Of this total, 49·3 million tons or about 26% were expected to come from captive or non-substitutable fuels, that is, fuels which are either produced by companies or organisations which themselves generated electricity, *i.e.*, Charbonnages de France, Ruhr, Belgian and Dutch coal producers; fuels which are the inevitable by-product of one or other industrial process, *i.e.*, blast-furnace gas; and fuels which can be cheaply and competitively produced and which are highly suited for purposes of power-generating, *i.e.*, lignite. Expressed in units of TWh, these non-substitutable 'fatal' fuels were expected to account for an output of 138 TWh by 1975, of which 98·9 TWh would be produced in Germany, which is the main producer of lignite in the Community area and where output from pithead power-stations was expected to continue to rise until at least about 1970. Similarly, in France and Belgium pithead power-stations are an important source of electric power and present levels of output were expected to continue throughout the whole of the period under review. From this, and given a total estimated demand for electricity in the Community area by 1975 of 789 TWh, it was estimated that the contribution from 'non-fatal' fuels would be of the order of 268·5 TWh in 1970 and 420 TWh by 1975.

Table 47. *Estimated fuel requirements of thermal power-stations in the six Common Market countries: 1960–75 (in million metric tons of coal equivalent)*

	Actual	Estimated		
	1960	1965	1970	1975
Belgium	6·3	7·3	9·1	11·2
France	13·7	22·3	31·5	43·8
Germany	42·1	55·8	72·0	90·0
Italy	3·1	9·6	18·0	24·5
Luxembourg	0·8	0·8	0·8	0·8
Netherlands	6·6	8·4	11·4	15·2

The final choice in determining whether new power-stations should be coal-fired, oil-fired or dual-fired would of course be governed by a number of different and sometimes conflicting factors. Out of all these, however, price was likely to be the main consideration and, in this connection, the Community experts quoted the following costs of investment of new plants according to the type of fuel being used:

Type of fuel used	Index of specific cost of investment
Good quality coal	100
Lignite and lower quality coals	105–110
Fuel oil	87–90
Natural gas	82–85
Dual-firing (coal/oil)	102–103

These figures were based on 150–300 MW plants. The Community experts then went on to calculate that the actual cost of investment according to the type of fuel used could be expected to be of the order of 150 dollars per kW for a coal-fired station, 132 dollars per kW for an oil-fired station and about 155 dollars per kW for a dual-fired station. As a result of this relative advantage in favour of oil, oil-fired stations were expected to increase somewhat more rapidly throughout the period under review. Thus, between 1960 and 1965, energy consumption by oil-fired stations in the Community area was expected to increase from 7·5 million tons of coal equivalent, or a 20·7% share of the substitutable part of the electricity generating market requirements, to 20·7 million tons, or 35·5% of the market requirements. Correspondingly, coal, although showing an increase in terms of absolute consumption from 28·7 million tons in 1960 to 37·6 million tons by 1965, was expected to show a fall in its relative share of the market from 79·3 to 64·5%.

In the 1966 study, the Community experts estimated that total consumption of electricity in the Community would reach 599 TWh in 1970 (that is, some 25 TWh more than in their original 1963 forecast) and 1214 TWh by 1980; in view of the existing forward construction programmes by the electricity industries in the six Common Market countries, it seemed probable that, out of the additional capacity of 180 TWh likely to come into operation by 1970, 30 TWh would be generated from new hydro, gas and lignite plants, 20–25 TWh from nuclear energy, and the remainder, that is, 125 TWh, from coal, oil and natural gas. Bearing in mind various national measures designed to secure a firm part of the market for indigenous coal production, the Community experts estimated that total coal requirements at power stations in 1970 could be expected to be between 66 and 79 million tons, with total oil requirements lying between 56 and 43 million tons of coal equivalent (depending to some extent upon the actual level of coal consumption).

Of the further increase of 600 TWh expected between 1970 and 1980, between 250 and 375 TWh seem likely to come from nuclear plants; between 325 and 200 TWh from conventional thermal plants and the balance of some 30 TWh from hydro and geothermal plants.

The 1966 study made no attempt to break down the individual fuel requirements by Community thermal power stations in 1980. It seems probable, however, that the vast bulk of any increase in fuel requirements at these plants will be met by oil and, possibly, some gas.

Conclusion

The sum of the energy requirements of the five major groupings examined in the Community experts' 1963 study, *i.e.*, the iron and steel industry, general industry, transport, the domestic sector and the power-generating sector, corresponds to a total energy demand figure for the Community area as a whole of 461 million tons of coal equivalent in 1960, rising to 570 million tons in 1965, 700 million tons in 1970 and 847 million tons by 1975—equivalent to an increase of some 50% in the first ten years and about 85% over the whole of the period. Differences in the rate of increase from one country to another are explained by variations in the rhythm of economic development, disparities in industrial and social structures (*i.e.* the very rapid rate of industrialisation which is currently taking place in Italy) and very big discrepancies in the rate of improvement in productivity.

Table 48. *Estimated gross energy requirements in the six Common Market countries: 1960–75* (*in million metric tons of coal equivalent*)

	Actual		Estimated	
	1960	1965	1970	1975
Belgium	33·9	41·6	42·0	48·0
France	121·9	154·7	187·0	231·0
Germany	205·3	251·0	282·0	330·0
Italy	65·6	100·8	137·0	176·0
Luxembourg	4·6	5·8	6·6	7·1
Netherlands	30·1	41·9	46·0	55·0
Total	461·4	595·8	700·6	847·1

Within these overall figures, demand for electricity was, as we have seen, expected to show a particularly rapid rate of increase. The 1963 study's anticipated comparative development and relationship between total demand for energy, G.N.P. and demand for electricity is set out in the graph opposite.

As a result it was estimated that *per capita* energy consumption in the Community area would rise from an average of 2,726 kg in 1960 to 3,250 kg in 1965 and 4,500 kg by 1975. It is significant that the 1975 figure for the Community as a whole—if realised—would still have been below the 1960 figures for the United Kingdom (4,800 kg) and the United States (8,200 kg). The exceptionally high figures for Luxembourg in the following Table 49 are explained by the fact that the Grand Duchy with its population

Comparative development of demand for energy, G.N.P. and electricity requirements in the Community area: 1960–75.

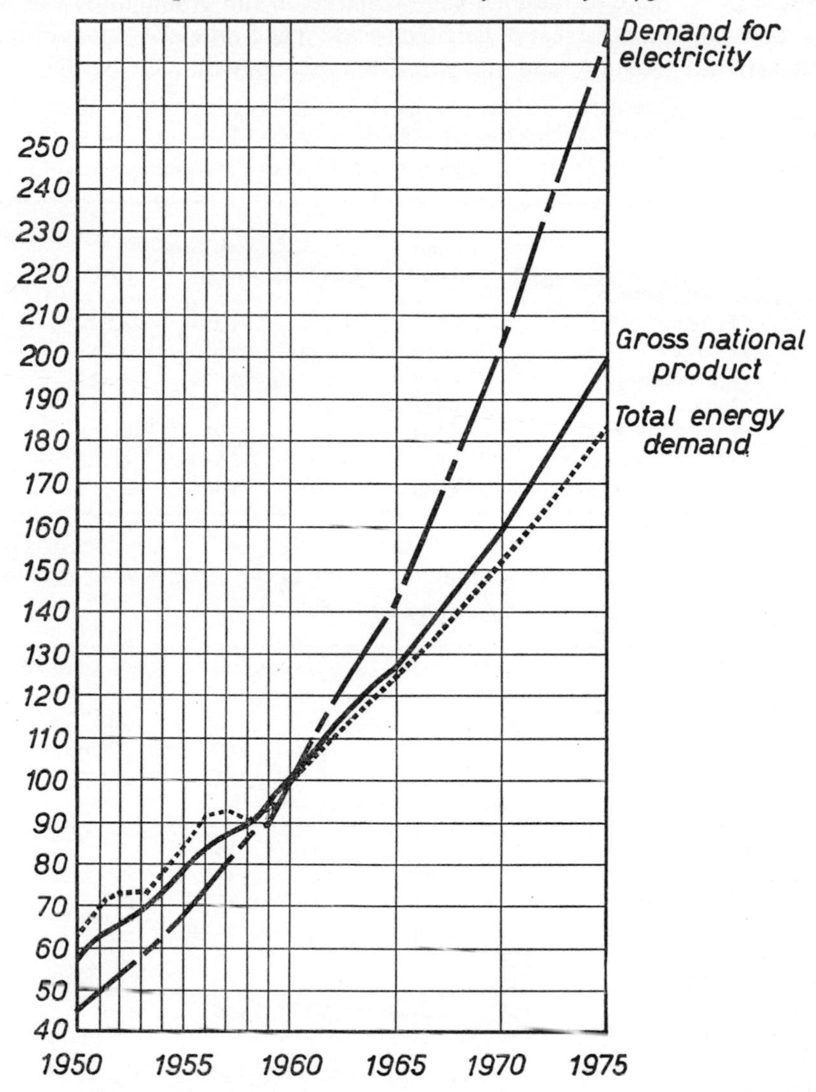

of only just over 300,000 possesses a modern and efficient steel industry which in 1960 produced over 4 million tons of steel.

The figures set out in Tables 48 and 49, while emphasizing the dramatic rise in energy consumption that had been taking place and was expected to continue in the Common Market countries, did not immediately reveal the very substantial shift in the relative levels of demand in each individual country. The biggest development was undoubtedly taking place in Italy, whose 9·1 % share of the total energy market in the Community area in 1950 was smaller than that of Belgium (9·86 %) and little above that of the Netherlands (6·96 %), and compared but poorly with that of the two

Table 49. *Estimated per capita consumption of energy in the six Common Market countries: 1960–75 (in kg/coal equivalent)*

	Actual 1960	Estimated 1965	Estimated 1970	Estimated 1975
Belgium	3,685	3,940	4,400	4,800
France	2,679	3,200	3,800	4,500
Germany	3,845	4,300	4,900	5,500
Italy	1,328	1,940	2,600	3,200
Luxembourg	15,333	18,770	19,700	20,700
Netherlands	2,617	3,150	3,600	4,100
Community area	2,726	3,250	3,800	4,500

Table 50. *Total energy consumption by sector in the six Common Market countries: 1965–80*

	1965 (actual)		1970 (estimated)		1980 (estimated)	
	(a) in million tons of coal equivalent	(b) as % of total	(a) in million tons of coal equivalent	(b) as % of total	(a) in million tons of coal equivalent	(b) as % of total
Steel Industry	61	10	66	9	74	7
General Industry	115	19	144	19	205	18
Transport	77	13	109	15	164	15
Domestic Sector	139	24	165	22	220	19
Electricity Generating	152	25	207	28	386	34
Energy Losses	52	9	52	7	81	7
Total	596	100	743	100	1130	100

giants, Germany, with 44·7 %, and France, with 28·5 %. By 1960, however, Italy had pushed into third place with 14·2 % and this was expected to rise further to 20·8 % by 1975. The relative shares of the market for each of the six individual countries are expected to be: Germany (38·9 %); France (27·2 %); Italy (20·8 %); Netherlands (6·6 %); Belgium (5.7 %) and Luxembourg (0·8 %).

By 1966, the Community experts in their later study estimated total energy requirements in the Common Market area in 1970 at 743 million tons of coal equivalent (*i.e.* 43 million tons or some 6% more than their earlier 1963 estimates); while total energy requirements by 1980 were expected to amount to 1130 million tons. These increases corresponded to a rise of 4·4% a year in energy requirements throughout the period under review—a figure only fractionally higher than that used in the original Long-term Energy Perspectives in 1963 for the period 1960–1975. In both studies (but even more markedly in the later one) the electricity generating and transport sectors provided the most rapid rates of increase in growth; between them they are largely responsible for the anticipated continuing rise in overall demand for energy.

The forward demand situation in the United Kingdom

Total energy consumption in the United Kingdom in 1965 was 288·7 million tons of coal equivalent. The National Plan[1] which was based on the assumption that G.N.P. would increase by 25% by 1970, estimated that demand for energy in the United Kingdom would rise from 286 million tons of coal equivalent in 1964 to between 324 and 337 million tons in 1970—this range being explained by the fact that in addition to a first estimate based on the forecasts made by each of the fuel industries with

Table 51.* *Fuel Industries Estimate of Inland Demand in the United Kingdom in 1970 (million tons of coal equivalent)*

	1964	1970
Coal		
For power-stations	68·0	84·0
For gasworks	20·5	10·0
For other purposes	98·7	81·0
Total	187·2	175·0
Oil (including petroleum gases)		
For power- stations	9·7	14·0
For gasworks	5·0	14·5
For other purposes	78·6	115·5
Total	93·3	144·0
Natural Gas	0·3	1·5
Nuclear and Hydro-electricity	5·1	16·5
Total inland demand for energy	285·9	337·0

* *Ibid.* p. 120.

[1] *The National Plan*, Cmnd. 2764, London, September 1965.

Table 52.* *Ministry of Power sector analysis of inland energy demand (million tons of coal equivalent)*

	1964	1970
Iron and steel	36·1	37·0
General industry	83·8	98·0
Railways	7·2	4·5
Other transport	29·1	38·5
Domestic	78·7	86·0
Other inland	50·4	59·0
	285·3	323·0
Stock changes and exports of coke and manufactured fuel	0·6	1·0
Total inland demand for energy	285·9	324·0

* *Ibid. p.* 120

regard to its possible level of sales in 1970 (and amounting in total to 337 million tons), the Ministry of Power made its own forecast based on its estimates of fuel consumption by each of the main consumer sectors, amounting to 324 million tons. For ease of reference, the two tables included in the National Plan are set out above.

The period of economic difficulty and uncertainty that followed the publication of the National Plan has, of course, called into question the assumptions on which it was based and, consequently, the likely forward demand for energy. Despite the absence of detailed sector forecasts and the present uncertainty about the rate of economic growth and energy in the short-term future, we have attempted to give some indication of the likely development in energy demand in each of the main consumption sectors in the United Kingdom, at least as far as 1970.

Iron and steel industry

According to the Special Report published by the Iron and Steel Board in 1964,[1] requirements of crude steel from home consumption in 1970 were expected to lie between 27 and 29·9 million tons (according to whether G.N.P. increased at a rate of 3 or 4% per year between 1961 and 1970). These figures were subsequently amended, however, in the light of the assumptions adopted in the National Plan, to give a total estimated demand in 1970 of 30·75 million tons of crude steel. The adoption of the

[1] Iron and Steel Board, Special Report 1964: 'Development in the Iron and Steel Industry'.

National Plan's forward estimates of demand would have required some slight amendment in an upward direction of the figures shown in Table 53 below, but these are of a negligible order.

Table 53. *Consumption of fuel in Iron and Steelmaking* (*in '000 tons*)

		1970 (estimated)	
	1960 (actual)	with 3% increase in GNP	with 4% increase in GNP
Non-coking coal	3,517	1,200	1,350
Coke	13,207	10,690	11,840
Coke breeze	1,528	2,380	2,640
Liquid fuels	3,348	4,550	5,050
Electricity (purchased million units)	6,083	10,100	11,200
Town gas (million therms)	139	160	180
Coke oven gas (million therms)	809	730	810

Source: Iron and Steel Board, Special Report 1964: 'Development in the Iron and Steel Industry', p. 85.

The report by the Iron and Steel Board further stated that the industry was likely to require some 18 million tons of coking coal to support the higher of the two rates of production postulated for 1970, compared with 16·9 million tons in 1963/64. Requirements of coals for purposes other than carbonisation during the same period were expected to fall from 2·2 million tons in 1963 to about 1½ million tons in 1970. The most rapid rates of increase in fuel consumption were expected to affect first, liquid fuels, consumption of which was expected to rise from 3·3 million tons in 1960 to between 4·5 and 5 million tons by 1970, secondly, electricity, demand for which was expected to increase from 8,327 million units in 1960 to 13,500 units by 1970 (including, however, about 2,500 units generated by the steel industry itself). Consumption of coke breeze was expected to increase from a level of about 2 million tons in 1963 to 2½ million tons in 1970. But the forward situation with regard to energy requirements in the iron and steel industry is complicated not only by the uncertain economic prospects for the next two or three years but also by the excess steel producing capacity in many parts of the world and the consequent pressure upon the United Kingdom market from overseas steel suppliers. These factors could well keep demand for home-produced steel by 1970 below 26 million tons with a consequent drop in energy requirements.

General Industry

Demand for energy in the general industrial sector amounted in 1960 to just over 77 million tons. While no detailed forward consumption estimates have been made for this sector, it was generally believed until quite recently that the greater part of the increase in demand would go to electricity. According to Mr J. E. H. Davies, at that time Managing Director of Shell-Mex and B.P., in a forward consumption estimate covering all oil products in the United Kingdom, which he made in the beginning of 1962,[1] deliveries of oil to this sector may be expected to rise to a level of 23¾ million tons (compared with just over 18½ million tons in 1963). This is probably a cautious estimate and it now seems likely that oil consumption in this sector by 1970 will exceed 25 million tons. Indeed, in view of the large refinery construction programme that has been put in hand in this country by the oil companies, this is one of the sectors, if not in fact the main one, on which they may be expected to concentrate their sales efforts for heavy fuel oil, both for new business and to displace coal from some of its existing markets. Since, to a considerable extent, oil will be displacing other forms of energy, the overall increase in demand for non-electrical energy in this sector is unlikely to exceed 2–3% a year. At the same time, the value of natural gas as a high quality heat source and feedstock may well result in a significant increase in its use by general industry if the early promise of the North Sea wells is maintained and translated into a substantial commercial availability at attractive prices.

Transport

Demand for energy in the transport sector in 1960 amounted to just over 30 million tons of coal equivalent. With the increase in air and motor traffic, demand in this sector may be expected to continue to rise at a comparatively fast rate, with a considerable element of substitution of diesel oil and electric traction in place of coal on the railways. An increase of some 6–7% a year leading to a total consumption figure of 40–42 million tons by 1970 may therefore be expected. According to Mr J. E. H. Davies, oil requirements in this sector in 1970 were likely to show the increase shown in Table 54.

In addition, demand for oil for ocean bunkers was expected to increase from 4·8 million tons in 1963 to 5 million tons by 1970.

The National Plan estimated total transport requirements by 1970 at 43 million tons of coal equivalent.

[1] See *Petroleum Times*, 6 April 1962, p. 233 for article by Mr Davies on the probable shape of consumption breakdown of petroleum products by the end of the decade.

Table 54. *Breakdown of oil consumption in the transport sector in the United Kingdom: 1963–70 (in '000 tons)*

	1963 (actual)	1970 (estimated)
Civil aviation fuels	2,238	4,350
Motor fuels	12,350	16,000
Railways	746	1,000
Total	15,334	21,350

Domestic sector

Demand for energy in the domestic sector in 1960 amounted to about 71·5 million tons (as in the case of the Common Market countries, the term 'domestic sector' includes households, local and Government services, handicrafts and agriculture). Despite the already high per capita level of consumption in this sector, and improvements in fuel utilisation, the growing popularity of central heating is expected to result in some further global increases in demand. At the same time there may well be, within the sector, some substitution, particularly in favour of electricity, oil and gas. Mr Davies, in his forecast, estimated that the use of oil for non-industrial heating purposes would rise to 9·7 million tons by 1970. Consumption of coal and coke may show a slight fall. Recently, there has been the prospect of natural gas from the North Sea and speculation on the effect that this new fuel may have on gross energy consumption in the domestic market. It may well be that natural gas will give a much-needed fillip to the trend towards higher domestic heating standards in this country with a consequent increase in the overall demand for energy. This will of course largely depend upon the price charged. In this event, the National Plan estimate of a demand figure by 1970 in this sector of 86 million tons may still be realised.

Power-Generating

This is the sector where by far the most rapid increase in demand is expected, rising from a level of some 64 million tons of coal equivalent in 1960 to well over 110 million tons by 1970. In 1960, over 90% of the electricity produced at conventional power-stations in the United Kingdom was derived from coal-fired plants. By 1970, it is estimated that this percentage will have fallen to about 84%, with oil accounting for a further 13%. The balance of 3% is expected to be dual-firing. On the assumption, therefore, that total fuel demand by the power-stations will amount to about 115 million tons of coal equivalent, including a contribution of some 16–17 million tons which will be met from nuclear plants, the National Plan estimated the coal requirements of the coal-fired stations at some 84 million tons a year, and oil demand from the oil-fired stations at 14 million tons a year. In view, however, of the recent decline in the annual rate of increase in electricity consumption, it now seems questionable whether these figures (and particularly those relating to coal) will be realised.

Conclusion

The absence of official detailed sector estimates of forward energy demand in the United Kingdom makes specific energy forecasts doubly difficult. The figures we have used in our brief review of consumption trends in the various sectors, as indeed in the case of the figures quoted in the National Plan, are therefore intended only as a guide, in order to facilitate comparisons between global trends in demand for energy by sector in the United Kingdom and continental countries and at the same time to give some indication of the change in the pattern of energy consumption in the United Kingdom which, while less dramatic and rapid than in many continental countries, is nevertheless taking place before our eyes at the present time.

There is no reason to suppose that in the longer term up to 1980 industrial activity and energy demand should not increase in the United Kingdom at much the same rate as in the leading industrialised countries in the European Community (of which Britain may very well be a member long before 1980). In the Community it has been estimated, as we have seen, that total energy requirements will rise at a rate of some 4·4% a year over the whole of the period 1965–1980. If allowance is made for the low rate of increase in the United Kingdom during the first few years of this period and we accept an average annual figure for the whole of the period of, say, 2·5–3·0%, then we arrive, for working purposes, at a gross demand figure in 1980 of some 415–450 million tons of coal equivalent.

The forward situation in Eastern Europe and the Soviet Union

As stated in the opening remarks of this section, there is very little detailed information available on forward energy requirements in the Soviet Union and Eastern European countries beyond 1965. Various national plans, while giving target figures for the production of various types of fuel and increases in the level of gross industrial output, rarely provide detailed or even adequate information about the development of energy requirements in individual sectors. The greater part of the meagre amounts of information that are available are only gleaned by means of careful sifting of press reports, industry bulletins or reports by international organisations, such as, for instance, the 1964 general report of the Coal Committee of the International Labour Office (ILO) in Geneva which stated that energy requirements in the Soviet Union were expected to increase by 120–130 % between 1960 and 1972. In the absence of detailed sector forecasts, however, and bearing in mind the rigid market disciplines exercised in the Soviet Union and the Eastern European countries generally over the production processes with a view to maintaining at all times a balance between supply and demand—except of course for those countries which are exporters of forms of primary energy (mainly in order to obtain Western currencies), *i.e.*, Poland which exports some 10 million tons of coal annually to the West, and the Soviet Union which exports varying quantities of coal and oil with Italy and France as her principal West European markets—the most reliable guide to total energy requirements in these countries by 1970 or 1980 probably lies in applying the historical rate of increase over the last ten years or so to the period under review:

Table 55. *Energy consumption in Eastern European countries and the Soviet Union: 1950–1965 (in million metric tons of coal equivalent)*

	1950	1955	1958	1960	1965
Bulgaria	3·3	5·7	7·8	11·0	18·0
Czechoslovakia	32·5	44·7	58·5	60·6	75·2
East Germany	49·5	70·5	74·9	79·6	91·6
Hungary	9·1	14·8	16·0	18·4	24·4
Poland	51·6	71·4	82·0	92·4	111·5
Roumania	9·9	16·7	19·5	25·1	41·9
Total	155·7	223·8	258·7	287·1	362·6
Soviet Union	272·3	429·2	576·5	668·9	850·1*

* The figure for the Soviet Union relates to 1964.

This shows that between 1955 and 1965 energy consumption in Eastern European countries rose from 224 to 363 million tons—an increase of 62 % —while in the Soviet Union there was an increase in energy consumption from 429 million tons to 850 million tons—a rise of 98 %. Within the global figures for Eastern Europe there are surprisingly small variations in increases in energy consumption from one country to another. Thus, although in Bulgaria energy consumption from 1955 to 1965 increased more than threefold, this is largely explained by the fact that this is the smallest and was the least industrialised of the six Eastern European States under review; in Roumania the figure for 1965 was more than two and a half times that for 1955, but in Czechoslovakia, Hungary and Poland the rate of increase lay in each case between 62 and 74%. East Germany with an increase of only 30% lagged well behind. These results would appear to support the argument that as a result of their political, ideological and general economic similarities, these six countries may be treated for purposes of examination of economic trends as an entity likely to show a comprehensive pattern of development. If, therefore, we take—in the absence of any more satisfactory information—as a crude working hypothesis the historical 1955–1965 average rate of increase in energy demand in these six countries and apply it to the remaining years of the period we are examining (*i.e.* up to 1980), we obtain a global energy demand for Eastern Europe by 1980 of about 650 million tons. If we envisage a high rate of increase in industrial activity, G.N.P. and, consequently, a rising annual energy demand in these countries from now until 1980 of, say, 5%, we arrive at a maximum total demand by 1980 of over 700 million tons of coal equivalent.

In the case of the Soviet Union the application of the historical rate of increase in energy demand to the period under review would give us a total demand figure by 1980 of over 2,000 million tons, which would give the Soviet Union a demand figure for energy approximately equivalent to the entire European O.E.C.D. area. An even more ambitious estimate, however, is the one provided by the forecast referred to above from the 1964 general report of the ILO. Under this hypothesis the energy requirements of the U.S.S.R. would amount by 1972 to between 1,458 and 1,518 million tons of coal equivalent, while the arithmetical extension of this rate of increase up to 1975 would result in a total energy demand by that year of some 1,800 million tons of coal equivalent.

The intention of the Soviet Union to drive ahead towards full-scale electrification of the country is well known and has been the subject of a number of papers and articles which are available to Western commentators.[1] Thus, Vikent'ev writing in the *Voprosy Ekonomiki* in 1963 stated that 'the grandiose tasks of the Soviet Union's development during the long-term period will require certain changes in the end use of national income . . . this presupposes, first of all, complete electrification of the

[1] Two of the most informative papers have been those by Sinev, Baturov and Schmelev on 'Trends of nuclear power development in the U.S.S.R.' given at the Third Geneva Conference in May 1964, and by A. Vikent'ev in the *Voprosy Ekonomiki* 1963, no. 1, on the subject of 'Prospects for the growth and utilisation of the national income of the U.S.S.R.'

country and, on this basis, improvement of the equipment, technology and organisation of production in all branches of the economy'. Turning from general objectives to more specific production targets Vikent'ev continued: 'Electrification of the economy plays the leading role in the development of all branches of production. Over the twenty years (*i.e.* 1960–80), construction of power installations in the U.S.S.R. will go ahead on a gigantic scale. It is intended to build 180 hydro-electric power stations, 200 large thermal electric power stations, and about 260 district heating stations. This will make it possible to increase electric power station capacity from 66·7 million kW in 1960 to 190–220 million kW in 1970, and to 540–600 million kW by 1980. At the same time, on the basis of accelerated technical progress in power engineering itself, the power base will be greatly increased, outstripping the development of the other branches of the economy. The long-term plan envisages that the new power capacity will operate on the highest technical foundation. According to the standard designs, and using prefabricated, reinforced concrete structures, we will build 600,000, 1,200,000 and 2,400,000 kW capacity thermal electric stations with generating units of 100,000, 150,000, 200,000 and 300,000 kW. Construction of super-powerful hydro-electric power stations will be carried out on an extensive scale'. Similarly, Sinev, Baturov and Schmelev, in their paper given at the Third Geneva Conference in 1964, declared that 'the creation of a material and technical basis for the communist society in the U.S.S.R., the development of industry and agriculture, and the increase of energy consumption due to a rising population, *demand* an expanded construction of electric power stations and a higher rate of power generation'. Power production and installed electric capacities in the U.S.S.R. were consequently planned to increase to 900–1,000 billion kWh by 1970

Year	Production of electrical energy in 10^9 kWh	Installed capacity in million kW
1945	43·3	11·1
1950	91·2	19·6
1955	170·2	37·2
1956	191·7	43·5
1957	209·7	48·4
1958	235·4	53·6
1959	265·1	59·3
1960	292·3	66·7
1961	327·0	73·9
1962	369·3	82·4
1963	411·6	92·4
1964	455·0 (estimated)	102·4
1965	508·0 (estimated)	113·0

(*i.e.* 180–200 million kW) and 2,700–3,000 billion kWh (540–600 million kW) by 1980.

For purposes of comparison, production figures for electricity (*i.e.* in terms of 10^9 kWh and installed capacity) in the Soviet Union from 1945 to 1965 are shown on p. 155.

Throughout this period the principal fuel used in power stations in the Soviet Union was coal although its share out of all fuel supplies was expected to fall from 70% in 1958 to just over 61% by 1965. A useful table showing the shares of the various types of fuel in the Soviet power-station market was included in an article in the *Revue Petrolière* in 1963[1]:

Type of fuel used	1958 (actual, %)	1965 (estimated, %)
Coal	69·9	61·3
Liquid fuels	5·9	12·3
Gas	10·7	17·2
Peat	8·1	5·7
Shales	0·9	1·2
Others	4·5	2·3
Total	100·0	100·0

Thus oil and natural gas were expected to show substantial increases in their share of the power-station market at the expense of coal and, to a lesser extent, of peat.

II. Availability

The questions of energy availability and demand have become fundamental factors in the advanced and mechanically and technologically sophisticated societies of Western Europe. Both these factors, however, contain an element of uncertainty. While demand for energy some five or ten years hence can only be calculated in relation to an estimated rate of growth in G.N.P., availability of energy consists of two related elements: the energy produced in Europe and the energy that has to be imported from extra-European sources. Europe's main indigenous source of energy, and one in which she is tremendously rich, is coal. All other sources, oil, natural gas, lignite, or hydro-electric sites, are small in terms of potential output of energy in relation to coal reserves. The problems facing the coal industry in Western Europe are not problems of shortage of supply, but of costs compared with the prices at which other forms of energy can be brought from points outside Europe to consumers within the Western European area. Imported energy, which has been increasing so rapidly during the

[1] *Revue Petrolière*, no. 1054; July–August 1963. See article by P. Empis: 'Le Gaz Naturel en U.R.S.S.', pp. 37–48.

whole of the post-war period, is at the moment in plentiful and cheap supply. New oil reserves, in particular, have been discovered within the last few years in the Middle East, Libya and North Africa and their exploitation has brought large new quantities of oil onto the nearby Western European markets. Whether this plentiful and cheap availability will prove to be a long-term feature of the European energy supply position, however, is open to doubt. Our next step will be to examine the availability of all sources of energy in Europe during the period under review.

Coal

It will be recalled that in its availability forecast of indigenous Western European coal supplies, made in 1960, the Robinson Commission estimated that production of hard coal would rise from a level of 477 million metric tons in 1955 to between 440 and 480 million tons by 1965 and 430 and 495 million tons by 1975:

	1965	1975
European Coal and Steel Community Countries	230–50	220–60
United Kingdom	205–20	200–25
Others	5–10	10–10

The pessimistic tone adopted by the Robinson Commission with regard to the future of the Western European coal industry was largely due to the general situation prevailing in the industry at the time when the report was written. The year 1959 witnessed the worst part of the crisis that faced the industry in the early post-Suez period and it was not until the end of 1960 that the situation began to show any sign of improvement. Nevertheless the view expressed at that time by the Robinson Commission reflected accurately the attitude of those circles which were fully in favour of a free trade policy in energy in Europe: 'It is inevitable that at the present, with very large stocks of unsold coal accumulating in almost all coalfields, the profitability of further investment in expanding coal production should be questioned. It is not easy to judge on the evidence as yet available the exact extent to which the immediate difficulties of the European coal industries are due respectively to decline in aggregate demand and to the substitution of oil for coal . . . There remains the fundamental problem of the future competitiveness of coal prices. Studies of the long-term relations of coal prices to other prices show that over the period since 1913 the relative price of coal, as compared with all other wholesale prices, has risen in all industrial countries . . . coal had by 1953 become some 50% more expensive than it had been in 1913 both in Europe and in America. But, while the relative prices of coal have continued to rise in Europe, the

American relative price has been constant or declining since 1946. Over the whole period the prices of crude oil has varied closely with the average of all other wholesale prices. Thus there has been an apparent long-term tendency for European coal to be less competitive with other fuels which has only in part if at all been compensated by more rapid increase in the efficiency of methods of using coal.'[1]

Six years later, in 1966, the O.E.C.D. Energy Committee in its report 'Energy Policy: Patterns and Objectives' expressed the view that only a continuing high level of protection would enable indigenous European coal production to be maintained at anything like present levels in the face of severe competition from oil, natural gas, imported coal and nuclear power. 'Some countries' the report continued 'may also have problems in finding the necessary coal-mining labour in areas of profitable production in view of the claims on manpower from other sectors of the economy. We have, therefore, assumed a further slight decline in production during the 1970s in order to arrive at the upper end of the range for 1980 and have taken a figure of 380 million tons.'

'Estimating the lower end of the range is more difficult. Apart from the uncertainties affecting the long-term trend of the costs of indigenous coal production and the prices of competing fuels, there is also a considerable area of flexibility in the future policies of the producing country governments. On the one hand, these governments will aim over the years at reducing the level of coal production. On the other hand, they will naturally strive to contain the rate of contraction of their coal industries within limits which will enable policies for redeployment and other social policies to operate successfully. It is also possible that the rate of contraction would in any case diminish as coal production becomes progressively concentrated on the most economic pits. Taking all these factors into account, we assume a possible figure for the lower end of the range for 1980 would be about 300 million tons. This would imply contraction by 1970 to some 375 million tons.'[2]

Like the Robinson Report, the 1966 O.E.C.D. Energy Committee Report stated that the delivered price of American coal in Western Europe could be expected to remain stable and might in fact decline if European countries were prepared to enter into long-term contracts, thus encouraging the building and use of giant colliers. (It should perhaps be recalled here that the Americans are full members of the O.E.C.D.; it would have been surprising, therefore, if they had not made the fullest possible use of the

[1] 'Towards a New Energy Pattern in Europe', *op. cit.* p. 42.

[2] 1966 O.E.C.D. Energy Committee Report, *op. cit.* pp. 86–87.

opportunities their membership affords them of presenting their coal export availability in the most favourable light.)

If the Robinson report in 1960 and the views expressed by the O.E.C.D. Energy Committee in 1966 were not of an optimistic nature as far as the future of the European coal industries was concerned, the voluminous study of the three European Executives on the long-term energy prospects of the Community was even more bleak. This pessimistic attitude was based primarily upon the assumption that coal would not be able to maintain a competitive position *vis-à-vis* imported fuels. Thus, while the price of imported energy was not expected to show much variation during the next ten or twelve years, indigenous coal prices were expected to rise substantially as a result of estimated wage increases of about 3½–4 % per year. With wages accounting for over 50 % of total costs in the coalmining industry, this was an exceptionally important factor. Increases in productivity were expected to be able to absorb wage increases in a number of the more economic of the Community coalfields (i.e. Ruhr, Aachen, Saar and Limbourg) until the middle of late 1960s; thereafter, however, rising costs were expected to outstrip the effect of increased productivity in all the coalfields of the Community. The pattern of increased underground output-per-manshift in the Community pits was expected to be as follows:

Table 56. *Estimated increases in underground output per man shift in the coal producing countries in the Community area: 1960–75* (*in kgm*)

	Actual	Estimated	
	1960	1965	1975
Ruhr (incl. Aachen)	2,185	2,700	3,750
Saar	2,055	2,700	3,700
Campine	1,790	2,350	3,200
South of Belgium	1,450	1,760	2,390
Nord/Pas-de-Calais	1,560	1,680	2,490
Lorraine	2,580	2,850	4,220
Limbourg	1,830	2,240	3,230

(It is only fair to point out that just as the Community experts in their original long-term estimates prepared in 1963 under-estimated the growth in, and demand for, energy in the Common Market countries, so they also miscalculated the likely growth in output per manshift in the pits: by the middle of 1966 output per manshift in Germany, for instance, had already reached 2,895 kg.)

The Community experts accordingly concluded that, faced with intense competition from other cheaper forms of energy, rising costs, growing difficulty in raising capital for new investments and a general feeling that coal was on the decline, the Community coal industries would inevitably contract in size and level of output unless they were artificially propped up by means of Government assistance. It is worth noting that any hesitations that appeared in the Community study about a drastic reduction in output of coal were due to social apprehensions; it was evidently believed that the near virtual dependence of the Community upon imported sources of energy did not pose grave security or balance of payments problems—the first could be solved by diversification in sources of supply, the second, by the satisfactory level of European gold and dollar reserves and the favourable trade outlook for the Community generally. As a result the Community experts estimated that only a substantial measure of Government support could prevent coal output in the Community falling to about 125 million tons by 1970–75. In other words any tonnage produced over and above this figure would be dependent on financial assistance by the public authorities. The amount of assistance required to maintain various levels of output between 125 and 225 million tons was estimated as follows:

	Disposable output			
Aid envisaged	1960	1965	1970	1975
(*a*) Aid of $2 with stabilisation of output after 1975	230	215	195	150
(*b*) Aid of $3 for period 1963–70 $2 after 1970, stabilisation of output after 1975	230	225	205	150
(*c*) Aid of $2 and stabilisation after 1980	230	215	195	165
(*d*) Aid of $3 for 1963–70 $2 after 1970, stabilisation of output after 1980	230	225	205	170

According to the amount of financial aid given, the level of Community coal production was estimated at between 125 and 225 million tons by 1970 and between 125 and 200 million tons by 1975. Similarly, coal imports were expected to vary between 30 and 110 million tons in 1970 and 40 and 100 million tons in 1975—their range being a factor of the level of Community coal production, itself determined by the amount of aid accorded.

By 1966, in their revised version of the long-term energy prospect for the Common Market countries,[1] the Community experts estimated coal consumption in 1970 at between 200 and 233 million tons, with the lower figure corresponding to a rate of decline broadly equal to that observed

[1] *Nouvelles réflections sur les perspectives énergetiques à long terme de la Communauté Européenne*, Luxembourg, 1966, *op. cit.* p. 32.

over the last few years, whereas the higher figure takes account of the various protective measures for coal that have recently been taken in Germany and France, particularly those designed to increase indigenous coal consumption at power-stations.

Coal imports into the Community, mainly from the United States, are expected to range between 32 and 35 million tons a year up to 1970. Assuming, therefore, that the Community experts' estimates of total coal consumption are correct, this, it is calculated, would leave a market for indigenous Community coal in 1970 of between 168 and 198 million tons a year. The Community experts emphasized the measures that have been taken by all four Community coal-producing countries to streamline their industries and to reduce the level of output and stated clearly that they expect this trend to continue until such time as the production of indigenous coal has been brought into line with effective demand. Effective demand in this instance should clearly be equated with the lower end of the range quoted by the Community experts.

Looking to the longer term, up to 1980, the Community experts envisage continuing and increasing pressure upon indigenous coal from natural gas, nuclear energy, oil and imported coal. Since only a few of the Community's coal-producing areas have the favourable geological conditions required for the application of remote-controlled mining, the Community experts foresee a steady decline in output to a total of between 100 and 185 million tons. Here, once again, it is the lower figure which they clearly believe to be the more probable and the more desirable. The Community's 1966 study of long-term prospects makes no separate estimate for coal imports by 1980 but this will obviously be a factor of Community coal production and the level of oil imports. However, if the Community were to decide to close down its coking-coal pits on a massive scale in order to enable the steel industry to import the major share of its fuel requirements from the United States, then coal imports could easily soar to between 60 and 75 million tons a year. The effect of such an increased import requirement upon price and availability is difficult to calculate. It is highly doubtful, however, whether the American coal producers would be able or prepared to maintain their present price levels for such a substantially enlarged requirement. Such a development in the Community would give them a virtual monopoly of coking coal supplies, and the temptation to increase prices, especially in the face of growing market difficulties in the United States, might prove difficult to resist. This is a point that is fully appreciated in many Community circles and constitutes in fact the most powerful argument for maintaining a substantial level of indigenous coal production. The realisation of the Community experts' forecasts with regard to forward coal production would mean that by 1970 indigenous coal output would cover only 23–27% of the Community's total energy requirements; other indigenous resources would provide a further 22% so that 50–55% of the Community's energy requirements would have to be imported from outside the Community. By 1980, indigenous coal production would account for only 9–16% of total energy requirements, while the gap between total indigenous energy resources and total demand might be as great as 62% or, if expressed in tons of coal equivalent, no less than 705 million tons—only 38 million tons less than total Community energy requirements from all sources in 1970.

Table 57. *Community Energy Balance for 1970 and 1980*

	Indigenous Community Production		Imports		Total	
	1970	1980	1970	1980	1970	1980
	(million tons of coal equivalent)					
Coal	168–98	100–85	32–35		200–33	
Lignite	36	40	2	695–525 (Coal, Lignite, Oil)	38	865–800 (Coal, Lignite, Oil)
Oil	28	30–50	370–337		398–365	
Natural gas	47	120–40	6	10–20	53	130–60
Hydro or Geothermal Energy	41	46	2	1	43	47
Nuclear Energy	11	90–125	—	—	11	90–125
Total	331–61	425–585	412–382	705–545	743	1,130
	(in % of total requirements)					
Coal	23–27	9–16	4–5		27–32	
Lignite	5	3	—	61–47 (Coal, Lignite, Oil)	5	76–70 (Coal, Lignite, Oil)
Oil	4	3–4	50–45		54–49	
Natural Gas	6	11–13	1	1–2	7	12–15
Hydro or Geothermal Energy	6	4	—	—	6	4
Nuclear Energy	1	8–11	—	—	1	8–11
Total	45–49	38–51	55–51	62–49	100	100

According to the 1966 Community study future coal production in the Common Market countries may be expected to decline at a very much faster rate than total requirements of solid fuel. If this hypothesis proves to be correct, then clearly the growing gap between coal demand and Community production of coal would have to be met by imports, mainly from the United States. In this connection it is interesting to observe that in their 1963 study of the long-term energy prospects of the Common Market countries, the Community experts, although giving a fairly wide range for possible prices of U.S. coal by 1975, stated firmly that they believed these prices were likely to be at the lower end of the range and would therefore remain fully competitive in West European markets:

Table 58. *C.i.f. prices of U.S. coal in W. Europe 1960 and 1975 (in $ per metric ton)*

	Coking Coal		Steam Coal	
	1960 (actual)	1975 (estimated)	1960 (actual)	1975 (estimated)
Pithead price	5·25–6·50	5·55–8·35	4·40	4·55–5·30
Transport costs in U.S.A.	4·50	4·40–4·64	4·50	4·40–4·65
∴ f.o.b. price	9·75–11·00	9·95–13·00	8·90	8·95–9·95
Atlantic freight	3·50	3·40–4·00	3·50	3·40–4·00
∴ c.i.f. price	13·25–14·50	13·35–17·00	12·40	12·35–13·95

More specifically, the price of U.S. steam-raising coals was not expected to exceed about $13·00–13·25 while the cheaper range of coking coals (a mixture of the high-quality Pocahontas and other coals) was not expected to rise beyond $13·35–15·05. Even the Pocahontas coals, while given a range of $15·40–17·00, were not expected to rise much above the lower level. The Community experts foresaw no difficulty about continuing to obtain substantial or increased quantities of U.S. coals, even though at least one of the studies on long-term availability of U.S. coal referred to by them[1] while forecasting a continuing export trade for U.S. coals to Japan, makes no reference to long-term exports of U.S. coals to Europe. No price increases were expected before 1965. Thereafter increased wages and costs were estimated to cause a possible increase in delivered prices of up to 10%, although it was believed that part of this could be absorbed by further progress in rationalisation by the two American railway companies, i.e., the Chesapeake and Ohio, and Norfolk and Western Railways, and lower transatlantic freight charges as a result of the introduction of larger 40,000 to 50,000 ton coal carriers. Thus early in 1965, ATIC, the French Government-sponsored and -controlled coal importing agency, commissioned the construction of a giant 85,000 ton coal carrier to ply regularly between Hampton Roads and Le Havre, and followed this up with a 60,000 tonner one year later. The 1966 study by the European Communities stated firmly that the prices for American coal exports from now

[1] Schurr and Netchert: *Energy in the American Economy 1950–75*. April 1960.

to 1970 (and beyond) could confidently be expected to remain at about $13–13·50 for mixed coking coals and $11–11·50 for electricity coals.

The theory that U.S. coals would continue to remain plentifully available at prices that could be relied upon to remain relatively stable was given considerable support by the report commissioned in 1963 by the U.S. Department of the Interior from a firm of American economic consultants on the future prospects for U.S. coal exports.[1] It should be noted that the background to this report was the heavy pressure being brought to bear upon the U.S. Administration both for social reasons to support the American coal industry and for foreign exchange reasons to endeavour to break down some of the foreign barriers to American coal imports with a view to promoting such exports.[2]

The Nathan report estimated the potential market for U.S. coal in Europe, divided into coking, or metallurgical, and other than coking coal by 1970, as follows:

Table 59. *Potential market for U.S. coals in Western Europe in 1970* (*in million metric tons*)

	Coking Coal		Steam Coal		Total	
	(*a*)	(*b*)	(*a*)	(*b*)	(*a*)	(*b*)
Belgium/Luxembourg	2·0	2·0	1·0	1·0	3·0	3·0
France	2·0	4·0	4·0	17·0	6·0	21·0
Germany	—	3·0	9·0	18·0	9·0	21·0
Italy	7·0	7·0	4·0	4·0	11·0	11·0
Netherlands	1·0	1·0	3·0	3·0	4·0	4·0
United Kingdom	4·0	15·0	—	7·5	4·0	22·5
Other European Countries	2·5	2·5	1·5	1·5	4·0	4·0
Total	18·5	34·5	22·5	52·0	41·0	86·5

(*a*) corresponds to the potential volume of imports from the United States by 1970 under policies in the importing countries placing 'moderately increased emphasis on low cost energy supplies'. It assumes a reduction in coal output in the Common Market countries by 1970 to a level of 195 million tons.

(*b*) corresponds to the potential volume of imports from the United States by 1970 in the event of the importing countries deciding to adopt policies 'placing the maximum degree of emphasis on a low cost energy policy that can reasonably be anticipated'. It assumes a reduction in coal output in the Common Market countries by 1970 to 164 million tons (as well as a subsidy of $2 per ton); while in the United Kingdom imports 'would exceed exports by about 10 million tons, with consequent effects on coal output'.

[1] Robert R. Nathan Associates: 'The Foreign Market Potential for United States Coal.'

[2] Robert R. Nathan Associates, vol. II, chapter 4: 'Given the availability of oil at prices competitive with coal or even lower, competition between coal and oil in substitutable

Turning next to prices, the Nathan report makes the following estimates with regard to the delivered prices of U.S. coal in Europe by 1970:

Table 60. *Estimated prices of U.S. coal in Europe in 1970*
(*in $ per metric ton*)

	Pocahontas Coking Coal	Mixed Coking Coal	Steam Coal
Pithead price	6·80–7·15	5·50–5·80	4·60–4·85
Railway charges	3·50–4·00	3·50–4·00	3·50–4·00
f.o.b. U.S. coast	10·30–11·15	9·00–9·80	8·10–8·85
Delivered price in Northern Europe			
Transatlantic freight	3·00–3·50	3·00–3·50	3·00–3·50
∴ c.i.f. price	13·30–14·65	12·00–13·30	11·10–12·35
Mediterranean area			
Transatlantic freight	3·00–4·00	3·50–4·00	3·50–4·00
∴ c.i.f. price	13·30–15·15	12·50–13·80	11·60–12·85

A comparison of these figures with those of the study by the European Communities, shows at once that the American figures are substantially more optimistic about the long-term trend in U.S. coal export prices.

Explaining how these prices had been arrived at, the report commented: 'It is difficult to predict future trends in mine prices. However, it is reasonable to expect that unit labour costs will probably not rise, that unit material costs may continue to rise with the price level, say by from 15 to 25 cents per metric ton between 1960 and 1970, and that other costs, including profits, may rise by from 10 to 25 cents per metric ton. The mine price for U.S. coal in 1970 might be expected to be from 25 to 50 cents per metric ton above the 1960 levels. The average value at the mine in 1960 was $4·73 per net ton or $5·20 per metric ton. An increase of 25 or 50 cents by 1970 would raise the average value by 5 or 10 per cent.'[1]

uses is much influenced by policies and actions by governments to protect their indigenous coal producers. This is a factor in most fuel-importing areas that are potential markets for U.S. coal, notably, the E.E.C., the United Kingdom, Canada and Japan. Because energy requirements have been, and are, rising so rapidly, these protective devices are operated mainly to reduce the competitive disadvantage of indigenous coal and to slow down somewhat the rate of penetration of the market by oil, though in some countries they have not been sufficient to prevent absolute declines in coal production in the past five years. National or regional decisions as to the degree of future reliance on indigenous coal, and the related protective measures invoked to effectuate them, will have a substantial, if not controlling, influence on coal-oil competition.'

[1] *Ibid.* vol. II, p. A15.

The Nathan Report also attempted to demonstrate that the level of American coal reserves was such that, despite a rapidly growing internal demand for coal—particularly in the electricity generating sector where coal requirements were expected to rise to between 700 and 750 million tons a year by 1975—continuity of supplies could be maintained in the long-term without any difficulty. To this end coal reserve figures were quoted of between 13,000 and 14,000 million tons, that is, recoverable at current costs, in the Appalachian Mountains while further reserves of up to 27,000 million tons were available at a production cost estimated at only 25 cents per ton above the 1960 rates. These figures were abstracted from a U.S. Geological Survey which stated that there were enough coal reserves in the United States to maintain the 1960 level of production for over 1,000 years. The Nathan Report quoted these figures as convincing proof that the American coal industry had the ability to meet both the anticipated rise in home demand and a high level of exports.

Against this overall background of plentiful availability of coal in the United States and restrictions on imports of U.S. coal in the majority of European countries the Nathan Report made a number of recommendations, some of purely commercial nature, but some also of a political and more controversial character. The full text of these recommendations is set out below:

'*Recommendation 1.*

That members of the coal industry, individually and jointly, make offers to major foreign consumers of coal of long-term contract at economic prices, with price provisions that limit increases to cost increases, and pass on to the purchaser the benefit of cost decreases.

Recommendation 2.

That the coal industry establish an association for the purpose of developing foreign coal markets, with authority to maintain offices abroad to represent the interests of members; to analyse and distribute information on foreign coal markets; trade restraints and competitive fuels; to conduct promotional and educational programs abroad; to represent its members in relations with the United States Government; to participate in international efforts to standardise coal testing and classification and to provide for inspection of coal export shipments.

Recommendation 3.

That the coal industry consider acquiring financial interests in coal distributing firms abroad; provide technical assistance to potential foreign consumers on the efficient use of United States coal; and determine the extent to which the coal industry should co-operate with foreign governments and consumers in the maintenance of foreign stock-piles of United States coal.

Recommendation 4.

That the United States Government make a careful examination of the potential contribution of expanded coal exports to the balance of payments and the domestic economy in relation to other trade objectives; determine whether current efforts to achieve a reduction of foreign coal import barriers are consistent with the relative priority warranted by this objective; and develop a program for the effective negotiation of a reduction of such foreign barriers.

Recommendation 5.

That the United States Government establish an Inter-Agency Coal Export Policy Committee, under the Chairmanship of the Secretary of the Interior, with representation from that Department, the Department of State, Commerce and Treasury, the Council of Economic Advisors and the Office of the President's Special Representatives for Trade Negotiations. The Committee would be responsible for systematic consultation with the coal industry and coal carrying railroads; co-ordination of government activity relating to the promotion of coal exports; and the conduct of studies required for the formulation of government policies and programs related to coal export expansion.

Recommendation 6.

That the United States Government establish a Coal Export Advisory Council composed of representatives from the anthracite and bituminous coal industries, merchant coal exporters, and the coal carrying railroads. The Council would be the institutional counterpart of the Inter-Agency Committee on Coal Export Policy and be responsible for the presentation to government of industry plans for the development of coal exports, consultation with all interested elements of the coal industry, and analysis of problems and recommendations for action by related industries such as the railroads.

Recommendation 7.

That the United States Government, the coal industry, and the coal carrying railroads concern themselves with the possibility of a reduction in rail rates on export coal. The Government should review the existing differential between rates on export coal and on shipments for domestic use, and should facilitate consideration by the railroads of special discounts for incremental movements. The coal industry should assume responsibility for representing the interests of foreign consumers in lower rail rates before the railroads and the I.C.C. The railroads should examine the importance of rail rates to the development of coal exports, and the possibility of passing on to foreign consumers the benefits of cost reductions resulting from expanded exports.'

In fact, up to the time that this book was written, few of these recommendations had been carried into effect, although there has been some pressure from American representatives at various international meetings and organisations for a more liberal attitude on the part of other Governments towards American coal imports. Even there, however, the American representatives have normally concentrated on developing the themes of the plentiful availability of American coal, its cheapness and the advantage to be gained by the purchasing country in negotiating long-term contracts. It was possible that more detailed discussions could have taken place in the Kennedy round negotiations in the GATT but with both the European Community and the United Kingdom having placed coal on the exceptions list, any dramatic developments in this respect were highly unlikely.

It is worth glancing for a moment, however, at the main arguments that have been advanced against excessive reliance on American coal imports into Western Europe: the steadily rising productivity of European mines; the irreversibility of pit closures, that is, capacity abandoned for the sake of the short-term price advantage of American coal; the growing proportion of American coal supplies that will have to be mined in areas further removed from the Atlantic seaboard with consequently increased inland transportation costs; the anticipated rise in inland demand for coal in the United States to a level of 800–900 million tons by the 1980s and the effect of this increase on coal reserves and the possible trend towards working less profitable seams; the question of whether American coal in the longer term will be able to remain competitive with oil and, except for certain specific consumers like the iron and steel industry, with nuclear energy. There are also political imponderables such as, for example, French objections to meeting more than a comparatively small proportion of their import requirements from the United States.

In the United Kingdom, the present stated production plans of the National Coal Board are aimed at maintaining an annual production capacity of about 180 million tons of coal. Whether it will prove possible to maintain sales (including exports) at anything near this figure will depend, first, on the assistance the Government is prepared to give the industry in its battle against other competing fuels, notably oil and, secondly, on the industry's own ability to streamline, to reduce costs and, above all, raise its productivity.

A decisive factor may well turn out to be the success or failure of the completely remotely-operated mines which the National Coal Board is striving to develop as quickly as possible. One such pit, Bevercotes, in the East Midlands Division of the N.C.B., has become the main testing ground of remote-control methods and techniques which have placed the British coalmining industry ahead of the rest of the world in the application of science to mining. At the originally modernised but conventional Bevercotes pit it was plannned to produce some 1¼ million tons of coal a year with a labour force of some 2,000 men. As a result of the decision to make Bevercotes into a remote-controlled pit, it is expected that by 1967 the colliery will be able to produce 1½ million tons a year with less than 800 men at an average output per manshift of 8 tons or more. Towards the middle of the next decade the British coalmining industry expects to produce about half of its output from remotely-operated faces. With a rate of productivity more than four times that of the average figure for today, the potential beneficial effect upon the overall position of coal in the energy market is evident. Admittedly, remote-control mining is still at the experimental stage. It is also true that these techniques are not suited to, or cannot be applied in, a number of coal-producing areas. A further and vital factor is, of course, the question whether the return of the financial expenditure involved will be justified. Nevertheless, the application of remote-control techniques to the coalmining industry holds out great promise for the future and must give pause to many who are inclined, rather too hastily, to write off coal as the energy source of the nineteenth century.

With coal production likely to decline steadily from its present level in most of the coal-producing countries in Western Europe, the already sharply different trends in energy production and consumption in Eastern and Western Europe will be further accentuated. For, although coal will show some decline in its relative share of the overall energy market in Eastern Europe, in absolute terms output is expected to continue to rise at a substantial rate. Thus, coal production (including lignite) was expected to rise from 513 million tons per year in 1960 to 612 million tons in 1965 (although coal's share of the total energy supplies of the Soviet Union was expected to fall during the same period from 57·1 to 45·3 %). Soviet coal production, including lignite, is expected to reach a level of 665–675 million tons in 1970 and 1,000 million tons by 1980. Within these estimates, production of coking coal was expected to increase from 110 million tons in 1960 to 150 million tons in 1965 (and 200 million tons by 1980).[1] In Poland, coal production has been rising at a rate of about 3 % per annum since the middle 1950s and was expected to reach a figure of 118 to 119 million tons by 1965; output of brown coal was expected to rise rapidly to 25 million tons by 1965 (compared with only 15 million tons as recently as 1963). By 1970, it is planned that hard coal production should reach a total of some 135 million tons with further increases thereafter. In Hungary, output of hard and brown coal is expected to show a modest increase to reach a level of 33–34 million tons by 1970 (compared with 31 million tons in 1963).

[1] The Soviet Seven Year Plan 1959–65 provided for a level of steel production of 90 million tons by 1965 and 250 million tons by 1980.

Lignite

Production of lignite in Western Europe has risen steadily since the turn of the present century. From a level of only 12 million tons in 1900, it has risen to 53 million tons in 1925 and 105 million tons in 1955. The bulk of this output has been in West Germany, which accounts for over 75% of all exploitable reserves in Western Europe. The Robinson Commission estimated the production of lignite would increase to 150 million tons in 1965 and 200 million tons by 1975, mainly for consumption by power-stations. Expressed in terms of hard coal equivalent, this would mean an output of 45 million tons in 1963 and 60 million tons in 1975. There was no separate estimate for lignite production in the 1966 O.E.C.D. Energy Committee Report.

The European Executives in their 1963 study of the Community's long-term energy requirements drew largely for their information upon a German study completed in 1962[1] and gave a production estimate of 117 million tons by 1975. Their subsequent 1966 study covering the period up to 1980 looked to a level of output of between 126 and 133 million tons in 1970 (equal to 36–38 million tons of coal), and 140 million tons in 1980 (equivalent to some 40 million tons of coal). There has, however, been a noticeable decline in the popularity of lignite over the last two years or so, especially in the domestic sector where it is widely used in the form of briquettes. This trend shows every sign of a structural decline and it is probable that in future the power-stations will provide virtually the only sizeable outlet for lignite production with some adverse effects upon proceeds.

The primary producers of lignite in Eastern Europe (other than the Soviet Union, which we have already considered above) are Eastern Germany and Czechoslovakia, where production in 1965 amounted to 251 million and 72 million tons respectively; Eastern Germany being the largest single producer of lignite in Europe. Plans are in hand in both these countries for substantial increases in production, although details of the precise levels it is hoped to achieve are not available. Thus, it was planned that production in Eastern Germany should reach an annual level of some 300 million tons by 1970, but recent reports that the East Germans, and indeed other East European producers, may now be reconsidering their planned increase in lignite production, have not so far been confirmed.

[1] *Untersuchung über die Entwicklung der gegenwärtigen und zukünftigen Struktur von Angebot und Nachfrage in der Energiewirtschaft der Bundesrepublik unter besonderer Berücksichtigung des Steinkohlenbergbaus*, Berlin, 1962.

Despite the fact that the known reserves are by far the biggest in Europe, it is only recently that plans for the large-scale exploitation of these deposits have been put in hand. This is largely due to the fact that until the end of World War II, East Germany was, of course, part of the Reich and constituted one of the richest farming areas in Europe with little traditional heavy industry. Furthermore, the ample supplies of hard coal available from Upper Silesia and the Ruhr were more than sufficient to meet the country's requirements. East Germany provides therefore an unusual example of a country which is almost entirely dependent for its fuel supplies upon lignite—in 1962 lignite accounted for only fractionally under 80 % of East Germany's primary energy requirements. Small increases in production of lignite are also expected in Bulgaria and Roumania.

Oil

Despite the post-war discoveries of crude oil in Austria, France, West Germany, Italy and the Netherlands, total production of crude oil in the Western European area remains comparatively small and is unlikely to show any great increase. The Robinson Commission forecast a production figure of 20 million tons in 1965 and 35 million tons in 1975 (equal respectively to 30 and 50 million tons of coal equivalent). The 1966 O.E.C.D. Report estimated indigenous oil production at between 25 and 30 million tons in 1970 and between 20 and 40 million tons by 1980. Clearly, by far the greater proportion of Western Europe's growing demand for oil will have to be met by imports.

Neither the O.E.C.D. Reports nor the European Executives in their studies anticipated any difficulties in meeting Europe's growing oil requirements. They pointed out that the two fundamental characteristics of the oil industry in the post-war period have been, first, to maintain a substantial surplus of installed capacity so that actual producibility is always considerably greater than production (or demand); and, secondly, to maintain the low technical cost of the marginal ton of production. As a result the establishment of a natural price equilibrium has proved an exceptionally difficult task.

The 1966 O.E.C.D. Energy Committee Report stated that, on the basis of present world proved reserves of 45–50 thousand million tons, there was enough oil to satisfy the current level of world demand for the next 30–40 years. The estimate, it was stated, took no account of likely improvements in methods of recovery which could well lead to a rate of recovery of 50 % compared with only 30 % at the present time. The Report concluded that 'Summing up and judging from past experience, there is evidence that sufficient further oil reserves will be developed to cope with any foreseeable increase of world demand well beyond the period under consideration'.[1] The actual level of imports was estimated at some 370 million tons in 1970 and 620 million tons by 1980.

The Community studies similarly emphasised the massive nature of world oil reserves;[2] the 1963 Report stating that total availability of crude oil was expected to increase from a level of 946 million tons in 1960 to between 1,195 and 1,245 million tons in 1965, between 1,475 and 1,585 million tons in 1970 and 1,795 and 1,900 million tons by 1975.

Area	Total reserves in 1965		1965 level of output expressed as a % of total reserves
	Billion tons	%	
Middle East	28·4	61·1	73·4
United States	5·0	10·7	11·7
Venezuela	2·4	5·2	13·5
Africa	2·5	5·4	32·9
N. America (excluding U.S.A.)	0·9	1·9	22·5
West Indies and S. America	1·2	2·6	21·3
W. Europe	0·4	0·9	20·0
Far East	1·6	3·4	50·3
Total	42·4	91·2	34·9
U.S.S.R. and Eastern block	4·1	8·8	16·6
Overall total	46·5	100	31·8

Turning to oil prices during the period under review, the 1963 Community study gave the following levels:

Table 61. *Cost of crude oil according to source* (*in* $ *per ton*)

	Production costs	Taxes and Royalties	Total costs
United States	18	4	22
Venezuela	7	7	14
Middle East	2·5	5·3	7·5
Sahara	9–11	(not known at that time)	

In their price estimates for 1975 the community experts assumed that the price of fuel oil would rise more rapidly than that of petrol or other refined products and worked on the hypothetical price of 17–19 dollars per

[1] 1966 O.E.C.D. Energy Committee Report, *op. cit.* p. 46.

[2] Estimated world petroleum reserves (*Source*: *Nouvelles réflexions sur les perspectives énergétiques à long terme de la Communauté Européenne*, *op. cit.* p. 54).

ton (this was made up of $10·5–12 for f.o.b. crude oil plus $5 for freight and $1·5 for refinery costs) at the Channel ports and $16–18 at Mediterranean ports. They freely admitted, however, that the eventual price was subject to a number of unknown factors, including the rate at which discoveries of new oilfields continued to be made, the volume of natural gas supplied to the European markets and its price, the policy of the United States and the Soviet Union with regard to production within their countries and their attitude towards exports, the final shape of the Community's common energy policy, and the development after 1970 of atomic energy for power-generating purposes.

The amount of fuel oil that can be produced from one ton of crude oil not only varies with the source of supply but is, furthermore, fairly elastic. Thus, in the United States, fuel oil accounts for only some 10 % of refined products while the proportion of light oils produced is considerably greater than in European refineries. The actual amounts of fuel oil or middle distillates produced can with modern technology be easily varied by means of cracking processes. Nevertheless the further development of modern technical processes is now running up against two problems: the first resulting from certain technological difficulties, the second from the tremendous growth in demand in Western Europe for fuel oil. In these circumstances the most satisfactory solution lies not in improving still further cracking techniques but in increasingly diversifying the sources of supply with a view to concentrating on imports of heavier crudes which are more suitable for the Western European oil markets. Broadly speaking the main technical problem facing European refineries in the next decade will be the maximisation of output of middle distillates and the minimisation of the production of petrol.

These technological factors apart (which they were confident would be resolved in the near future), the Community experts did not envisage any difficulties in the field of oil supplies up to 1970. Favourable prices were considered to be assured as a result of the big post war discoveries, the rapid increase in oil output in the Middle East, the world-wide ramifications of the big oil companies and their consequent ability to regulate prices, the vast expansion in refinery-building programmes and the development of oil pipelines. Under these circumstances demand and availability of oil in the Community area was confidently expected to reach a minimum level of 365 million tons (that is, about 510 million tons of coal equivalent) by 1970 and might well exceed 600 million tons (or about 840 million tons of coal equivalent) by 1980, thus accounting for over 60 % of the Community's energy supplies by that date.

The 1963 Community study had estimated that the overall pattern of oil consumption in the Community area by 1975 would be as follows:

	Actual	Estimated		
	1960	1965	1970	1975
Oil for internal energy consumption				
Motor fuel	30	48	65	80
Refineries own use	7	14	17	21
Total specific uses	37	62	82	101
Fuel oil and other energy products	49	87–89	132–150	166–201
Oil for other requirements				
Bunkers	11	15	19	23
Oil for non-energy purposes	7	12	17	22
Total oil requirements	104	176–178	250–268	312–347

The British Government's White Paper of October 1965 on Fuel Policy[1] took much the same line as the O.E.C.D. and Community Reports on the forward availability of oil: 'There is little doubt that there are adequate reserves of oil in the ground to meet rapidly growing consumption throughout the world . . . looking ahead, the industry is confident that additional reserves will be proved as needed . . . In addition, there are vast reserves, estimated at seven times the proved oil reserves of the non-communist world, in the shale oil deposits in the United States and in the Athabasca and South American tar sands'.[2] The White Paper did, however, sound a note of warning that, even with a fairly plentiful supply of oil, there was a chance that, eventually, prices would rise as costs of exploration and recovery had to increase.

The policy of the British Government, as it appeared from the 1965 White Paper, is broadly to ensure sufficient refinery capacity within the United Kingdom to meet the likely level of demand: 'The Government has, however, adopted the objective that home refining should suffice to cover inland demand and bunkers in total. This should help to ensure that the proceeds of product exports are generally more than sufficient to cover the cost of product imports.'[3] In fact, refinery capacity by 1970 was expected to amount to not less than 102 million tons, while the oil industry's estimates for production and demand for 1970 totalled 97 and 93 million tons respectively. This suggested that the Government's objective would be realised at least up to 1970. Given, however, the rapidly rising demand for oil, the forward refinery construction programme will, presumably, have to be kept continuously under review.

[1] Cmnd. 2798. [2] *Ibid.* p. 20. [3] *Ibid.* p. 22.

Information with regard to the forward production of oil in the Soviet Union and Eastern Europe is once again incomplete. The 1958–1965 Soviet Seven Year Plan provided for a level of oil output by 1965 of 340 million tons of coal equivalent (compared with 211 million tons in 1960). Actual production in 1965 was 243 million tons of oil, or just under 350 million tons of coal equivalent. Production was expected to rise further to between 690 and 710 million tons by 1980. The Eighth Five Year Plan accordingly provided for a production of 345–355 million tons of oil or over 600 million tons of coal equivalent by 1970. In Eastern Europe, only Roumania is an oil producing country of any importance, with reserves estimated at 135 million tons of oil. In the case of the Soviet Union, realisation of these plans would mean an annual rate of increase in production between 1961 and 1980 of the order of 8%—compared with a rate of increase of only 4·3–4·4% for coal. This increase in oil production is expected to occur primarily in European Russia. This in turn is expected to lead to a substantial increase in the size of the Soviet oil pipeline network, and to result in increases in exports of Russian oil, in particular to East Germany, Czechoslovakia and the Baltic area. (According to a recent study by the United States Office of Oil and Gas the Soviet Union was expected to be in a position to export some 70 million tons of crude oil by 1965 and this figure was expected to rise to 80 million tons by 1970).

While, however, there is no reason as yet to doubt that oil will remain available in adequate quantities throughout the period up to 1980, there must be a considerable measure of doubt as to whether the oil companies will be able to maintain the present level of prices. While it is true that oil reserves over the world as a whole still amount to more than 30 years' consumption at the current level, it is significant that the ratio between reserves and current consumption has been falling steadily in recent years (i.e. from over 37 years in 1960 to only just over 30 years in 1965). It is also a well-known fact that oil companies go in for a large measure of self-financing and must find from their profits the growing capital sums needed to meet the cost of exploration, transport and refinery construction. According to a statement made by Mr Ashton, the Treasurer of Esso, a year or two ago, the world oil industry outside the Communist block was faced with the task of finding some £95,000 million of capital over the next 15 years. This is equivalent to £6,300 million a year and was to include an increase in capital investment of some £2,500 million a year. Such a rise inevitably requires higher profits and these in turn can only be got from higher prices.

Hydro-power

The Robinson Commission in 1960 forecast an expansion in hydro-electric capacity from a net production figure of 140 TWh in 1955 to 240 TWh in 1965 and 350 TWh by 1975, corresponding to 56, 95 and 140 million tons of coal equivalent respectively. Commenting on these figures the report stated that 'hydro-power occupies a well-determined place in the overall electricity supply. The run-of-river plants, though their output varies seasonally, are of particular value for covering base-load. Storage hydro plants are most satisfactorily used to meet a daily peak load, or, where storage capacity is large, a seasonal peak. Storage plants with pumping facilities are likely to become increasingly important for dealing with daily peak demands, and to be integrated for that purpose into a general system of electricity supply. In view of these particular roles, the development of electricity production from hydro resources will be little affected by slight possible variations in the rate of increase of total energy demand'. The Robinson Commission did not believe that the emergence of nuclear power would have the effect of holding back the development of hydro plants; indeed, they saw the latter as a useful supplement to nuclear power which is most suited to meeting base-load requirements. The main problem was therefore likely to be one of capital, particularly since many of the remaining hydro resources in Europe, and especially those in France, Italy, Norway and Sweden, are in the more remote and less populated areas of those countries. The Robinson Commission's estimates corresponded to an annual rate of increase in hydro-electric capacity of about 5½% between 1955 and 1965 and a little less than 4% from 1966 to 1975 and compared with an annual rate of increase during the first post-war decade of some 7%. If realised, more than two-thirds of Western Europe's economically workable hydro potential will have been harnessed by 1975.

The European Executives in their study, after stressing the very great differences in the availability of this source of energy in the six Common Market countries, stated that the economically exploitable hydro potential of two of the three largest members, that is, Italy and Germany, had already been largely harnessed. Of all the Common Market countries, only France offered much scope for an expansion of hydro-power since it was estimated that less than 50% of the country's hydro potential had yet been tapped (and this figure could be reduced to 40% if potential tidal-power resources were taken into account). As in the case of the Robinson Report, the study by the European Executives stressed the contribution made by hydro plants towards meeting fluctuating seasonal demands. The proportion of hydro plants geared to meet such seasonal demands in 1960 was put at 17% in Germany, 48% in France and 38% in Italy. By 1975, these proportions were expected to have increased to 20–25% in Germany and to between 50 and 60% in both France and Italy. Total production of electricity from hydro plants between 1960 and 1975 was expected to develop as follows:

Table 62. *Estimated production of hydro-electricity in the six Common Market countries: 1960–1975*

A. Output per country (in gross TWh)	Actual		Estimated	
	1960	1965	1970	1975
Belgium	0·2	0·3	0·3	0·3
France	40·9	46·9	51·0	61/55
Germany	13·0	15·4	19·5	21·0
Italy	48·2	43·2	57·5	63·0
Luxembourg	—	0·9	1·4	1·4
Netherlands	—	—	—	0·2
Total	102·3	106·7	129·7	146·9/140·9
In million tons of coal equivalent:	41·0	42·7	51·9	58·8–56·4
B. Proportion of electricity produced from hydro sources in relation to total electricity production				
Belgium	1·3	1·2	1·1	0·9
France	54·9	44·2	34·5	28·3/25·5
Germany	11·1	8·9	8·9	6·9
Italy	85·8	52·2	48·6	38·9
Luxembourg	—	39·7	41·2	40·0
Netherlands	—	—	—	0·5
Total Community area:	36·5	25·7	23·2	19·2/18·4

These figures show that despite an anticipated increase of some 40–45% in the amount of electricity produced by hydro plants in the Community area between 1960 and 1975, proportionately, the share of hydro-power in the overall level of electricity production was falling rapidly, that is, from 36·5 in 1960 to less than 20% by 1975, with a particularly rapid rate of proportional decline in Italy and France. These figures were based on the assumption of average hydraulicity and it is worth noting that, expressed in terms of kWh, the difference between a year of good or bad hydraulicity is now estimated at some ± 12 milliard kWh for the Community area. As in the case of other sources of energy, the production forecasts for indigenous sources of energy prepared by the European executives were based on the hypothesis that the landed c.i.f. price of imported energy during the period under review would remain at, say, $12–13 per ton of coal equivalent in Genoa. A substantially higher c.i.f. price could therefore have the effect of accelerating the development of hydro power, at least up to 1970; after that date, the near exhaustion of suitable sites for hydro-plants in the Community area will inevitably result in a slowing down in the rate of construction of new hydro capacity.

The 1966 study estimated a level of production from hydro-plants of only 120·1 TWh in 1970 and 150·7 TWh by 1980 (*i.e.* the latter figure is thus only fractionally above the 1963 study's forecast for 1975, due mainly to revised and less optimistic forecasts for Italy and Germany). Hydro-power's contribution towards meeting total Community energy requirements was thus not expected to exceed 6% in 1970 or 4% by 1980.

The most dramatic developments in the building of new hydro-stations in recent years have taken place in the Soviet Union, where the most powerful hydro-electric power stations in the world are known to be under construction. These include such giant plants as the Bratskaya hydro-electric station on the Angara river in Siberia which by the beginning of 1964 had an installed capacity of 3,600 MW (and is due to reach 4,500 MW); the Krasnoyarskaya power station on the Yenisei, also in Siberia, planned to reach 6,000 MW; the Lenen Volga plant with 2,300 MW and the Volgskaya plant with an installed capacity of 2,530 MW. While these giants are still in the planning and construction phase, by the end of 1965 the Soviet Union planned to have in operation eleven stations of over 1,000 MW. Output of electricity from hydro plants in the Soviet Union in 1962 was equal to just under 30 million tons of coal equivalent and accounted for less than 7% of total electricity production. The hitherto untapped potential of hydro-power, however, is enormous, as shown by the size of the stations now being planned or built. Indeed, the economically workable hydro-power resources of the Soviet Union exceed those of any other single country and are estimated by the Russians themselves at 2,100 billion kWh per year. Of these reserves, less than 15% are in European Russia, the remainder being found by the great rivers, that is, Angara, Yenisei, etc. in Siberia. It is expected that electrical energy produced from hydro-plants will increase by some three- to fourfold between 1965 and 1980. It is interesting to note however that in their paper to the Third Geneva Conference, Nekrasov and Schevinsky stated that it would be 'only the use of the most economical hydro-power resources on the basis of which large hydro-electric power stations can be built with construction costs per kilowatt of installed capacity on the level of thermal electric power stations and with the cost of generated electrical energy at 0·03–0·04 kopecks/kWh' will enable this target to be met. The same paper also confirmed that 'the hydro-power resources of the European districts of the U.S.S.R. are not only relatively small and cannot play an important part in the balance of electric energy supply of these districts but are also less effective economically'.[1]

Elsewhere in Eastern Europe considerable expansion programmes are planned or under construction in Poland, Czechoslovakia and Roumania. Thus, hydro-plant producibility is expected to increase by about 3,500 billion kWh in Poland, about 1,000 billion kWh in Czechoslovakia and about 1,200 kWh in Roumania.

Natural Gas

The Robinson Commission in their report estimated that production of natural gas in the European O.E.C.D. area would rise from 5,000 million m^3 in 1955 to 20,000 million m^3 in 1965 and 40–45,000 million m^3 by 1975, corresponding to 7, 25 and 50 to 60 million tons of coal equivalent respectively. Commenting on these figures, the Commission stated that 'the recent geological survey of O.E.E.C. countries, made under the aegis of the Oil Committee, clearly indicated that geological conditions in many Member countries give good prospects for finding new reserves of crude oil and natural gas . . . it seems likely that the limiting factors will prove to be the problems of finding markets and of building up long-distance transportation networks rather than the limits of actual reserves of natural gas itself'. While this has proved to be the case, both the French at Lacq and the Dutch in Groningen have been very successful in laying down new pipelines in order to secure, rapidly and effectively, satisfactory outlets for their natural gas, and this problem will probably have been overcome entirely by the late 1960s.

The 1966 O.E.C.D. Energy Committee's Report emphasised the difficulties of attempting to make any long-term forecasts of indigenous production, referring, *inter alia*, to the 13 years of drilling that were necessary before the fields at Groningen were eventually proved. The Committee's final estimate was one of 40–60,000 million m^3 in 1970 and 80–120,000 million m^3 by 1980, corresponding to between 50 and 70 million tons and 114 and 170 million tons of coal equivalent respectively. 'To achieve the lower figures would involve some additions to reserves with normal rate of

[1] Third Geneva Conference: 'Progress of Power Engineering and Atomic Power Plants in the U.S.S.R.' by Nekrasov and Schevinsky, pp. 7–8.

development, while the higher would involve substantial additions to reserves and very rapid development of them'.[1]

The experts of the European Executives in their study estimated the proved reserves of the Community area in 1962 at between 560 and 857 milliard m³; France—between 130 and 255 milliard m³; Germany—between 25 and 42 milliard m³; Italy—between 105 and 160 milliard m³; Netherlands—between 300 and 400 milliard m³.

These reserves were widely dispersed: thus, while in France all the proved reserves were known to be in the Lacq area, those in Italy were divided between the Po Valley (about 75%) and Calabria and Sicily; and those in Germany between the Ems/Dollard area (about 80%) and the deep south. Since this estimate was made both the official and unofficial estimates of the total value of the Dutch reserves have very much increased. Thus, as we saw in the previous Section, while the last official estimates put the Groningen reserves at 1,700 milliard m³, it is confidently believed in informed circles that the real value of the reserves exceeds this figure by a substantial margin.

Estimates of production costs in the Community area vary from 0·08 to 0·02 U.S. cents per 1,000 kcals. In view of the subsequent rise of the estimated level of reserves in Groningen—which is known to be cheap to produce—we may assume that the average cost is more likely to be of the order of 0·04–0·02 U.S. cents. The Community experts anticipated a rise in indigenous production of natural gas from 10·7 milliard m³ in 1960 to 16·8 milliard m³ in 1965, to between 24·2 and 27·2 milliard m³ in 1970 and to between 32·6 and 42·1 milliard m³ by 1975. In fact, Dutch output alone by 1970 could, it was stated, exceed 25 milliard m³ and might well reach 30–34 milliard m³ by 1975, while in Germany there might similarly be increases of between 25 and 50% over the estimates prepared by the Community experts as a result of recent discoveries. Taking these developments into account, the overall pattern of production for the Community area during the period was expected to show the development shown in Table 63.

The Community study also considered the likely development of imported methane by submarine pipeline or methane-tankers from North Africa and elsewhere. Interest in Algerian natural gas deposits was particularly strong in Community circles both for reasons of its geographical proximity and French anxiety at that time to secure privileged outlets in the Community for Algerian oil and natural gas. This element of French policy weakened considerably towards the end of 1963 however as relations

[1] 1966 O.E.C.D. Energy Committee Report, *op. cit.* p. 88.

between France and Algeria, particularly with regard to control and exploitation of the oil and natural gas deposits which had largely been developed as a result of French initiatives and French capital, became more clouded. The position today therefore appears to be that, as a result of political uncertainties, as well as a lack of a will to overcome the considerable technical problems involved by the construction of a submarine pipeline and the new discoveries of natural gas in the Netherlands, a large-scale flow of natural gas at an early date from the Sahara to the Common Market countries is unlikely. In their study, the Community experts estimated the probable c.i.f. price of natural gas at the French or Italian Mediterranean coast at about $2 per million kilocalories or $14 per ton of coal equivalent. (This was made up of some 20–35 % wellhead or production costs, 45–50 % transport costs both overland from the wellhead to the Algerian coast and transport by tanker or submarine pipeline to the European coastline, and about 30 % for stocking and distribution costs).

Table 63. *Estimated production of natural gas in the Common Market countries: 1960–75 (in milliard m³)*

	1960	1965	1970	1975
Belgium	0·1	0·1	0·1	0·1
France	3·1	5·5	6·0	7·5
Germany	0·7	1·7	3·5	5·5
Italy	6·5	7·5	8·6	10·0
Netherlands	0·3	2·0	20·0	25·0
Total	10·7	16·8	38·2	48·1
In million tons of coal equivalent:	13·8	21·7	50·3	63·3

In Northern Europe, that is, the Netherlands area, the delivered price of the natural gas was estimated at about $21 per ton of coal equivalent. Finally, the Community experts estimated that imports of natural gas into the Community area from all sources, but mainly of course from Algeria, might reach 6–9 milliard m³ by 1970 (8–12 million tons of coal equivalent) and 15–20 milliard m³ (20–26 million tons of coal equivalent) by 1975.

The more recent Community study of forward energy requirements up to 1980 has clearly taken into account the likely impact of the Groningen natural gas find, as well as the recent discoveries of natural gas in the North Sea. Forward consumption of natural gas in the Community countries in 1970 and 1980 is now expected to be as follows:

Table 64. *Natural Gas Consumption in the Common Market Countries: 1970–1980 (in thousand million m³)*

	1970	1980
Belgium	4·0	10–12
France	11·0	27–33
Germany	15·0	37–44
Italy	10·0	22–28
Netherlands	8·0	22–28
Total	48·0	118–145
In million tons of coal equivalent	52·9	130–60

Of the 130–160 million tons of coal equivalent envisaged in 1980, between 120 and 140 million tons are expected to come from the gasfields in (or just off the coasts of) the Common Market countries. The estimates, if realised, would give natural gas approximately 10% of the energy market of the Community.

It is tempting to speculate on the lessons that can be drawn for Britain from Community experience in the development and use of natural gas. Everything depends of course upon the size of the North Sea gasfields. The discovery of a gasfield of equal importance to Slochteren, producing by, say, 1975 some 50–75 million tons of coal equivalent, would undoubtedly have a significant impact upon the energy balance of the United Kingdom. Assuming that total energy consumption in the United Kingdom by 1975 will be of the order of 385 million tons of coal equivalent, this would give natural gas (on the basis of a production of 50 million tons of coal equivalent) a share of some 13%—a percentage not dissimilar from that forecast for the Community in 1980. It would, however, give natural gas a dominating role in making additional requirements and, bearing in mind the growing contribution from nuclear power stations, would mean severe pressure upon oil and coal markets. A limited find of some 12–20 million tons of coal equivalent, on the other hand, could be expected to have comparatively little effect upon established coal and oil markets. It is reasonable to assume that a substantial proportion of any natural gas discovered in the North Sea will be used to promote the economic development of essentially agricultural areas like East Anglia and Lincolnshire and to assist in the industrial development of Humberside. Natural gas could also make a notable contribution to Britain's balance of payments. Unless, however, the size of the North Sea gasfield turns out to be of vast proportions—and a very great deal larger than finds made up to the present time (October 1966) would lead us to suppose—the impact of natural gas on the British economy and energy scene must be seen in perspective and recognised as being a potentially valuable, but nevertheless modest, contribution to this island's soaring demand for energy.

In the Soviet Union production of natural gas has been rising dramatically throughout the post-war period. Thus, production increased from just under 7½ million tons of coal equivalent in 1950 to 62 million tons in 1960 and over nearly 142 million tons in 1964.

The natural gas industry in the Soviet Union has obviously expanded at a tremendous rate in the last ten or twelve years. Before the Second World War natural gas was practically unknown in the Soviet Union and the first major discovery was not made until 1943 at Elchanska, near Saratov.

In his report to the XXII Congress of the Communist Party of the U.S.S.R. M. Nikita Khrushchev gave some details of the target figures for the production of natural gas in the Soviet Union over the period from 1960 to 1980: this showed that production was expected to rise to between 310 and 325 milliard m^3 by 1970 and between 680 and 720 milliard m^3 in 1980, that is, an increase over the whole of the period of 1,500%. The Eighth Russian Five-Year Plan, however, covering the period 1966–1970, provided for an output of only 225–240 milliard m^3 by 1970 and suggested that the figures referred to by Mr Khrushchev a few years ago had proved unattainable.

P. Empis in his article on natural gas in the Soviet Union[1] had the following estimated breakdown of consumption of natural gas, by consumer, in the Soviet Union in 1965:

User	Quantity of gas taken in milliard m^3	Quantity of gas taken in %
Local usage	14·0	9·3
Use of gas as raw material	8·5	5·7
Use by industry	60·0	40·0
Metallurgical purposes	26·7	17·9
Cement industry	8·4	5·6
Machine tools	12·0	8·0
Other industrial purposes	12·9	8·5
Heat purposes	56·0	37·3
Pipeline requirements and losses	11·5	7·7
	150·0	100·0

[1] *Revue Petrollère*, No. 1054, July–August 1963: 'Le Gaz Naturel en U.S.S.R.' by P. Empis, p. 38.

Elsewhere in Eastern Europe, output of natural gas rose almost sixfold from 30,000 thousand million kcal in 1950 to nearly 170,000 thousand million kcal in 1965. Virtually the whole of this increase, however, was accounted for by Roumania.

With its rich Transylvanian deposits Roumania was until recently ranked as first in Europe (excluding the Soviet Union) in the size of her natural gas deposits (she has only recently been displaced by the Netherlands). In 1950 reserves of natural gas were estimated at 575,000 million m^3. Moreover, Roumanian natural gas has a methane content of 98–99% and is, in terms of calorific value, one of the best in the world. As in the case of the Soviet Union, natural gas, despite its high calorific value and exceptionally low cost, was scarcely used before the advent of the Second World War. The first deposits were in fact located at Sarmasel as early as 1907 and exploitation on a very minor scale was commenced in 1911/1912. Its industrial application began in 1917 with the construction of a 55 km pipeline from Sarmasel to Turda and iron founderies, cement works and glass and chemical factories were constructed along its route. It is expected that production will continue to increase at a steady, if unspectacular, rate of some 3–4% a year. In the other countries of Eastern Europe, natural gas is of comparatively little importance as a source of energy and no major developments are anticipated.

Nuclear energy

As we have seen, the Robinson Commission were unable to agree on a firm figure for nuclear energy production by 1970 or 1975 owing to the wide range of opinions and views with regard to the time at which nuclear power stations were likely to become competitive with conventional plants. The Commission consequently restricted itself to giving an upper and lower range between which it expected production to remain. On the former assumption, under which new construction in Europe in the period under review would be restricted to one new 500 MW capacity plant per year, installed nuclear capacity for electricity production would have been about 5 GW in 1965 and 10 GW in 1975. In this case nuclear development would have tended to continue, as the Report states, on a research rather than a commercial scale. Equally, on this basis, nuclear plants would have represented less than 4% of the total electrical generating capacity added during the period. The Commission took as an upper limit the likely level of installed nuclear capacity by 1975 in the event of nuclear energy becoming competitive by the late 1960s and the greater part of new additions to base-load capacity taking the form of nuclear plant from that time onwards. Under these circumstances installed nuclear capacity might have risen from 7 GW in 1965 to 35 GW by 1975. Assuming therefore a utilisation rate of about 6,500 hours in 1965 and 6,000 hours in 1975 and with load factors of approximately 75 and 70% respectively, the electricity production was estimated to be between 33 and 46 TWh in 1965 and between 60 and 210 TWh by 1975; expressed in terms of coal equivalent, the possible contribution of nuclear energy towards meeting primary fuel requirements in the European O.E.C.D. area was estimated to be between 15 and 20 million tons in 1965 and 30 and 90 million tons by 1975.

The 1966 O.E.C.D. Energy Committee report estimated installed nuclear capacity by 1975 at 40 GW (*i.e.* only 5 GW more than the Robinson Commission's estimate for the same year), but envisaged a more than twofold increase to 90 GW by 1980 as a result of the proposals or intentions of a number of Governments to embark upon substantial nuclear power station building programmes. Translated into tons of coal equivalent, this would mean an equivalent amount of nuclear heat released of some 170 million tons in 1975 and 440 million tons by 1980. While looking at the whole subject of the contribution to Europe's energy requirement from nuclear energy in a very broad light, the report's conclusion on this point is of interest: '. . . it is clear that nuclear power will make a great and growing contribution to energy supplies in the O.E.C.D. area during the period under review. The development and proving of this new source of energy on the scale envisaged will thus ease the heavy dependence on fossil fuels, especially in Western Europe, and introduce a new flexibility into energy policy. It will tend to set a ceiling to the prices at which fossil fuels can be sold for thermal production without substantial subsidy on protection. However, if nuclear energy is to keep this promise of becoming a new and practically inexhaustible source of cheap energy, it will be essential in the future that the research and development efforts supported by industry, national authorities and international organisations be maintained and improved.'[1]

The Community experts, in their 1963 long-term study of the Community's energy requirements, emphasised the high cost of nuclear development but justified this on the grounds of the need to gain experience of the running of nuclear plants and the contribution that nuclear energy could make to meeting the Community's long-term energy requirements. These costs were considered to be divisible into three main categories: charges arising from the immobilisation of capital, the cost of the re-

[1] 1966 O.E.C.D. Energy Committee Report, *op. cit.* p. 63.

fuelling cycle, and thirdly, exploitation and maintenance costs; the following estimate of costs for 1965–67 made by Euratom was quoted:

For enriched uranium	
installation costs	200–50 $/kW
refuelling cycle costs	2·4–3·5 mills/kWh
maintenance costs	5–8 $/installed kW
For natural uranium	
installation costs	250–80 $/kW
refuelling cycle costs	2·1–2·5 mills/kWh
maintenance costs	4–7 $/installed kW

But the Community experts also estimated that by 1968–70 the installation costs for a nuclear power station burning enriched uranium would have been cut down to 175 $/kW and that of the refuelling cycle to 2 mills/kWh or even less; while for a plant burning natural uranium further technical progress would probably result in a reduction in installation costs to below a figure of 250 $/kW and an estimated costs for the refuelling cycle of between 1·6 and 2 mills per kWh. It is worth noting that these estimates were less optimistic than those of the United States Atomic Energy Commission which in a statement made on 25 June 1962 declared itself to be of the opinion that it should be possible, for a nuclear plant of 600 MW, to reduce installation and building costs between 1966 and 1970 to between 135 and 170 $/kW and the cost of the refuelling cycle to between 1·5–1·8 mills per kWh.

Turning to the competitiveness of nuclear plants with conventional power stations the Community experts stated quite categorically that, assuming a price for coal or oil of $13 per ton of coal equivalent and a utilisation rate for the nuclear power plants of 6,000 hours per year, nuclear power-stations would be competitive in some parts of the Community by between 1965 and 1967 and in almost all parts of the Community by 1967–70. In this attitude, they were very close to the position adopted by the authors of the fourth and fifth French national plans which, as we have already seen, have consistently argued that break-even point for nuclear power plants would be achieved by the end of the 1960s. Indeed, the main difficulty foreseen by the Community experts in general, but by Euratom in particular, was not that of attaining break-even point by 1970—which they considered to be more or less assured—but of building up a sufficiently large force of trained atomic engineers to be able to put into practice the large-scale construction programmes of atomic power stations which they confidently predicted for the 1970s and 1980s.

As a result the Community experts estimated an increase in installed nuclear capacity in the Community area from 73 MW at the end of 1961 to 175 MW at the end of 1962, 700 MW at the end of 1963 and 1,000 MW at the end of 1964. By 1970 production of electricity from nuclear power stations was expected to reach 20–25 TWh (compared with 0·1 TWh in 1960). Thereafter, a much larger degree of uncertainty was admitted. Nevertheless, despite their optimism and obvious desire to set in motion an ambitious nuclear plant construction programme over the whole of the Community area, the Community experts rejected as excessively ambitious and technically impracticable the range given by UNIPEDE in 1962 of between 11,000 and 26,000 MW, corresponding to a production of electricity of between 65 and 150 TWh. Instead, the Community experts estimated electricity production from nuclear power plants in the Community area at 6·5 TWh in 1965 and 60 to 100 TWh by 1975:

Table 65. *Estimated electricity production from nuclear power stations in the six Common Market countries: 1965–75 (in TWh)*

	1965	1970	1975
Belgium	0·1	0·5	1–5
France	2·5	10–12	19–30
Germany	0·2	4–6	19–30
Italy	3·7	6–8	20–30
Netherlands	0	0·3	1–5
Total	6·5	20–25	60–100

It was further anticipated that the majority of the nuclear power plants would be constructed in areas devoid of conventional energy resources. Nor was any difficulty foreseen in maintaining adequate supplies of uranium since available proved reserves in Canada, Australia, South Africa and the United States—and exploitable at a cost of between 8 and $10 per lb of concentrated U^3O_8—were estimated at 600,000 tons. Even so, as the report states, the effect of the price of concentrated uranium upon the overall production cost per nuclear kilowatt hour is relatively slight (*i.e.* ± 0·4 mill/kWh or some 5% for reactors in current use). As far as supplies of enriched uranium are concerned, the report stated that, although there was no enrichment plant in the Community area and the United States held therefore a virtual monopoly in its supply, the existing agreements between the United States and Euratom or the individual European Governments guaranteed to European reactors the same degree of access as to American plants. Furthermore, the U.S. Atomic Energy Commission had stated publicly that it was prepared to enter into a number of long-term contracts with European consumers. In this connection it is worth noting that the existing American production capacity of enriched uranium was at that time sufficient to supply requirements of reactors totalling some 40,000 MW; a total which the Western world was unlikely to attain before the early years of the next decade. The controversial problem of security of nuclear supplies would be further resolved by the results of current research on methods to improve reactor performance. Thus improvements in the rate of thermodynamic efficiency would result in a greater degree of valorisation of the uranium input; while the recycling of plutonium and its use in breeder reactors as well as the use of thorium and the stocks of 'poorer' quality uranium would result in a substantial increase in the level of nuclear fuels. As a result of these measures it should be possible eventually to achieve a 50 % utilisation rate of the energy released by fission, using natural uranium, against the then current rate of less than 1 %. High stocks constituted another important factor in obtaining security of supply and here the burden of costs was very much smaller in the case of nuclear plants than in conventional plants (*i.e.* the Community experts estimated the cost of stocking of fuel as three to four times less in the case of a nuclear power plant). Comparative fuel and stocking costs were estimated as follows:

Table 66. *Comparison of estimated fuel and stocking costs at nuclear and conventional power stations*

			Nuclear plants	
	Coal-fired plants	Oil-fired plants	Natural Uranium	Enriched Uranium
Basic price of fuel (in $)	14 $/time	20 $/time	8 $/lb	
Increase in cost of fossil fuels as a result of stocking (in dollars per metric ton)	3–3·5	3–3·5	—	—
Resultant increase in cost of electricity (in mills/kWh)	1·0	1·0	0·1	0·2

It must however be noted that these estimates do not include building costs (which are very much higher in the case of nuclear plants) or maintenance costs. The Community experts concluded that on the basis of the information available the contribution of nuclear energy to assuring the Community's security of energy supplies was clearly demonstrated. This was because[1]:

[1] *Etude sur les Perspectives Energétiques à long terme de la Communauté Européenne, op. cit.* p. 116.

(*a*) The actual level of reserves of natural uranium in politically stable countries was such as to guarantee a regular supply sufficient to meet the requirements of an extensive and widespread reactor construction programme;

(*b*) Supplies of enriched uranium were guaranteed by the United States Atomic Energy Commission which was prepared to enter into long-term supply contracts. Existing enrichment plants were adequate to meet any foreseeable demand in the medium-long term;

(*c*) the cost of stocking nuclear fuels for power stations amounted to only a third or a quarter of the cost of stocking conventional fuels;

(*d*) In the light of current technical progress energy produced as a result of nuclear fission would be expected to provide in the longer term virtually limitless supplies, thus making a major contribution towards security of supply.

In April 1966 the Euratom Commission published an 'indicative' report[1] setting out the views of the Commission on the probable development of nuclear power in the Community from 1970 as far ahead as the year 2000. After briefly listing the various arguments in favour of a rapid development in nuclear power output (*i.e.* security of supply, cheapness and rational uses of all forms of energy) the report examined the likely demand for electricity in the Community up to the year 2000 (in thousand million kWh):

1960	1970	1980	1990	2000
272	575	1,080	1,930	3,450

Nuclear power's share in meeting new, that is, additional, demands of electricity was expected to amount to 40% between 1970 and 1980, to 60% between 1980 and 1990 and to 80% between 1990 and 2000. Even these proportions were regarded as constituting a minimum working hypothesis. On this basis, nuclear power would, by the year 2000, account for more than two-thirds of total electricity production and for 30% of total energy consumption in the Community.

The report gave the following figures as the minimum objectives for nuclear power production during the period under review:

Years	MW	Annual utilisation rate (in hours)	Net annual production (in 10^9 kWh)
1970	4,000	7,000	28
1975	17,000	7,000	120
1980	40,000	7,000	280
1985	78,000	6,900	540
1990	135,000	6,800	920
1995	226,000	6,650	1,500
2000	370,000	6,550	2,400

[1] '*Premier programme indicatif pour la Communauté Européenne de l'energie atomique*', *Journal official des Communaités Européennes*. Brussels, April 1966.

The report reviewed four possible construction programmes for nuclear power stations between 1970 and 2000. Programme I would rely exclusively on the continued construction of 'proved' nuclear power stations, *i.e.*, of the graphite and boiling-water type. Programme II envisaged, as from 1975 onwards, the introduction of advanced reactors whose share of the total would then increase progressively. Programme III differed from II only in that the Community would rely on fast breeder reactors as from 1980 instead of advanced reactors as from 1975. Finally, Programme IV was a combination of II and III; advanced reactors would take the place of graphite and boiling-water types as from 1975, but to these would be added, from 1980, the fast breeder reactors. Under Programme IV, fast breeder reactors would account by the year 2000 for 50% of all nuclear power, with advanced type reactors providing 30% and the graphite and boiling-water types 20%. The report concluded that Programme IV was by far the most attractive because of its immense technical, industrial and economic advantages.

Turning to prices, the report estimated the probable costs of conventional fuels for use in power stations as follows:

	$ per ton of coal equivalent
Community coal	14·59–18·42 at the pithead
Imported coal	11–12 c.i.f.
Heavy fuel oil	10·50–14·00 ex-refinery

The comparative costs of a building programme of conventional and nuclear power stations (the latter according to Programme IV above) for the period 1970 to 2000 were given as follows (the price of coal being taken as $12 at the pithead):

	(in $ thousand million)		
	Conventional Power Stations	Nuclear Power Stations	Difference
Construction costs	16·3	27·0	+10·7
Fuel costs	61·7	16·4	−45·3
Running costs	7·2	9·7	+2·5
Total	85·22	53·1	−32·1

In order therefore to be able to compete with nuclear power stations, as developed under Programme IV, it would be necessary for conventional plants to be able to buy their coal at a price of $5·7 per ton.

But the year 2000 is still a long way away and the type of programme envisaged by the Euratom Commission still faces vast problems and difficulties before its implementation can become a reality.[1] The forward estimates for nuclear power contained in the 'Nouvelles Réflexions sur les Perspectives Energétiques à long terme de la Communauté Européenne' also published in 1966 looked only as far as 1980. Here, the Community experts expressed the view that the original objective of 40,000 MW of installed nuclear capacity by 1980 would be substantially exceeded and might in fact reach 60,000 MW.

As far as the development of nuclear power in the United Kingdom is concerned, the Second Nuclear Power Programme provided for a total installed nuclear capacity of 13,000 MW by 1975. This figure however disguises a trend towards building steadily larger nuclear power plants: thus, the last plant being constructed under the First Nuclear Power Programme, at Wylfa, which is not expected to be completed till 1969, will have an installed capacity of 1,180 MW, while Dungeness B, expected to come into service in 1971, will have an installed capacity of 1,200 MW. These stations will, it is claimed, be able to generate electricity at substantially lower costs than conventional coal or oil burning plants: 'total generating costs from the second Dungeness station are estimated, on cautious assumptions, to be about 0·46*d.*/kWh. This calculation is based on the discount rate of 7½% used by the Central Electricity Generating Board for the appraisal of investment, and on a 75% load factor and a 20-year life. However, all components have been designed to have a life of 30 years at 85% load factor, and these more optimistic assumptions would lead to an estimated cost of about 0·38*d.*/kWh. These figures can be compared with costs from the coal-fired stations under construction or planned at Drax (4,000 MW) and Cottam (2,000 MW) of 0·52*d.*/kWh and 0·54*d.*/kWh respectively and from the oil-fired station under construction at Pembroke (2,000 MW) of 0·52*d.*/kWh or 0·41*d.*/kWh without tax'.[2]

As we have already seen, a similar trend towards larger nuclear power stations can be observed in the Soviet Union: 'The increase of atomic power plant capacity must necessarily be based on the increase of the unit capacity of reactors and turbogenerators. The installation of reactors with a capacity increase of between 1·5 and 2 times will also substantially reduce specific capital investments. Taking this tendency into account, work on the construction of new power units with a reactor capacity up to 350–400 MW has begun at Beloyarskaya and Novo-Voronezhskaya atomic

[1] It should perhaps be added that while the estimates up to 1980 set out in Euratom Commission's indicative report were based on firm plans by the six member Governments, the projections up to the year 2000 were prepared solely by the Commission.

[2] White Paper on Fuel Policy, Cmnd. 2798, *op. cit.* pp. 23–24.

power stations. Scientific and technical prerequisites for the construction of reactors with significantly larger capacity and atomic power plants with a capacity of 1,000 MW and higher have been created and specific design and research work is being conducted in these directions'.[1] Work in the nuclear field in the Soviet Union up to 1970 is expected to be concentrated upon research work into the development and construction of large scale prototypes, with a view to an increasing serial construction of large nuclear power plants in the decade 1970–80. Thus, by 1980, the total electrical capacity of nuclear power plants in the Soviet Union may be expected to amount to 'several tens of millions of kW'.[2]

III. Conclusion

The immediate conclusion to be drawn from an examination of the developments in the field of energy supply and demand in Europe since the end of the Second World War is that we have moved from a situation in which the economy of almost the whole of Europe depended upon one single source of energy—coal, to what can fairly be described as the beginnings of a multi-fuel economy. Today coal and oil together provide over 80% of Europe's total energy requirements, though it seems certain that within the next few years oil will, in fact, have displaced coal as the principal source of energy in Western Europe as a whole. Already the forward energy estimates for the six Common Market countries show that by 1970 oil consumption is likely to rise to a total of some 365 million tons of coal equivalent, thus providing about 49% of total energy requirements in the Community area and displacing coal as the principal source of energy; while by 1980 oil consumption could reach over 840 million tons

[1] 'Progress of Power Engineering and Atomic Power Plants in the U.S.S.R.', paper by Nekrasov and Sclevinsky at the Third Geneva Conference, 1964.

[2] See 'Trends of Nuclear Power Development in the U.S.S.R.' by Sinev, Baturov and Shmelov, Third Geneva Conference, 1964, pp. 6 and 13: 'In the U.S.S.R. work on further improvements of nuclear power reactors is mainly directed towards the possibilities of a major increase in reactor unit capacities of up to 1,000 MW and beyond and the development of optimal converters which will provide for the most rapid possible introduction of U238 into the fuel cycle. The design and development work carried out in the Soviet Union confirms, on the basis of the available experimental data, the technical possibility of constructing a uranium-graphite reactor cooled by water at supercritical conditions or gas-cooled reactors with unit capacities of the order of 1,000 MWe as well as fast sodium reactors with a breeding ratio of 1·7 with the breeding in the core close to 1. The steam conditions at nuclear power plants with the above reactor types will approach the conditions that can be achieved in modern conventional power stations. The above unit capacities would provide a reduction in specific capital costs and operational costs and secure the production of competitive power.'

of coal equivalent or more than 60% of total energy requirements. In the United Kingdom and Eastern Europe oil consumption is, similarly, rising at an exceptionally rapid rate, although its relative share of the overall energy market is likely to remain substantially below that of the Common Market countries, at least in the foreseeable future. In addition to coal and oil, however, hydro-power, natural gas, and of course nuclear energy, are becoming increasingly important as potential sources of energy, so that by the commencement of the next decade Europe will have moved into a situation which will be characterised by the emergence of a truly multi-fuel economy.

Desirable though this situation may be when considered in the long term and in relation to Europe's comparative poverty in terms of indigenous energy supplies, the situation that has developed in the energy market, particularly in Western Europe, bears within it, nonetheless, the seeds of a number of social and economic difficulties. The tremendous development of oil since 1945 from a situation where it was regarded as a useful supplementary source of energy to one where it is now successfully challenging coal as the principal source of supply, has posed dramatic problems for the Western European coalmining industries. The oil industry, with the advantage of world-wide ramifications, has been able to follow a marketing policy tailored to meet the prevailing mood or situation in any particular country or groups of countries; above all, it has enjoyed the valuable asset of the very large and easily exploitable oilfields that have been discovered and successfully developed since the war. Faced with this dynamic competition, the labour-intensive coal industries of Western Europe, often suffering from long periods of under-capitalisation, ravaged by war, and inevitably compelled during the war years, and for some time after the war, to give priority to achieving maximum levels of output with little provision for future production, have been forced onto the defensive. In all countries a large number of pits have been closed, mechanisation introduced on a large scale, and within the last two or three years plans made and put into operation for the application of automation to the industry. Thus, in the United Kingdom, Germany, and France, plans are well advanced for mining by means of remotely operated longwall faces. The first of these faces was put into operation in the United Kingdom early in 1963, and has demonstrated that it is practicable to operate a power loaded longwall face, including both the operation of the power loader and the movement of the armoured flexible conveyor and the mechanised roof supports, from a point some distance away and without any miners being required on the coalface.

Towards the end of 1966 Bevercotes colliery in the East Midlands Division of the National Coal Board, the first fully automatic coal mine in the world, came into production and by the end of 1968 the Coal Board expect to have a total of 20 remotely-operated faces in operation. Plans are also in hand for the early introduction of similar remote-controlled techniques in Germany and France. Despite the progress that has been achieved through increasing productivity, improving the quality of coal by better preparation techniques, and the modernisation of methods of transporting and delivering coal, the prices of oil in many Western European countries are sometimes so low that it is completely impossible for the coal industry to match them. This may sometimes be due to the fact that many independent small oil companies, which are able to operate without the large overheads of the big 'majors', are in a position to throw on to the market substantial quantities of cheap oil which tend to establish, or at least strongly influence, the general market price level; but the essential factor today when all has been said and done, is simply that, generally speaking, oil can be put onto the market at cheaper prices than coal. As a result it has proved necessary in almost all Western European countries for Governments to intervene and take action to prevent a situation developing in which their coal industries would have been faced with drastic closure programmes that entailed severe social, as well as economic, consequences.

Under these circumstances, the battle that has been waged between coal and oil has had deep emotional, social, economic and even strategic undertones. In all the countries affected by this evolution, pressure has been brought to bear upon Governments to take action to preserve the coalmining industry in order to guarantee the miners their livelihood, or at least to ensure that they would find alternative employment if the mine in which they worked was closed down. By and large, the social problem has not proved too difficult. That this has been so is due almost entirely to the fact that the period of recession in the European coal-mining industry has coincided with an era of high economic activity and general prosperity. Thus, the demand for manpower has remained at a high level and it has proved possible, particularly in countries such as Belgium, Germany and France, where the main concentration of heavy and medium industries are near the pits, to absorb, rapidly and effectively, any miners made redundant by pit closures. In the United Kingdom, where the older, and consequently exhausted or less economic, pits are usually in areas where the level of unemployment is already above the national average, the problem has been a much more difficult one and has only been resolved successfully by means of close and constant collaboration between management and the Unions. If, however, for the time being at least, the immediate social difficulties appear to have been surmounted, the general economic and strategic difficulties still remain.

The strategic element of the energy debate has, of course, resided in the security of supply factor. The coal producers of Western Europe, in their 1962 memorandum, 'Meeting Europe's Energy Requirements', stated that 'if by 1975 Europe's coal output were to be cut down to 300 million tons (against the current level of about 430 million tons), Europe would be compelled to import no less than fifty per cent of her energy needs'. (They also argued that if the expansion of the world economy continued at its present rate, then in ten to twenty years' time a fuel shortage would be much more likely than a fuel surplus.) By 1964 the six Common Market countries were able to provide from their own indigenous energy supplies no more than 57% of their total energy requirements. With indigenous production likely to remain static or to decrease, it is evident that imported energy will steadily become more important. This will result in a heavy drain on foreign exchange, and above all, place Western Europe in a position of dangerous dependence upon the rest of the world, and notably the oil exporting countries of the Middle East, for her energy supplies. It is the recognition of this security or strategical factor, more than any other single consideration, that has been at the basis of the great energy debate that has been waged in Western Europe during the last ten years. Recently, however, there has been a notable swing in the trend of the debate, under the impulse of natural gas discoveries and, above all, the growing conviction that nuclear power will become fully competitive before the end of the next decade, so that early protectionist policies, which were widely prevalent in Western Europe, have come under increasing attacks. Accordingly, the real debate in the Common Market countries has already become one of deciding not at what level coal production in the Community should be maintained, but which measures should be taken in order to maintain a reasonable level of production and thus avoid increasing still further the energy dependence of Europe upon the rest of the world. The individual Common Market countries, in fact, appear to have decided that their coal industries should cut back their production to a level of somewhere between 150 and 175 million tons; within this figure, particular attention, and assistance, will probably be given with a view to maintaining coking coal capacity, especially if, as it has been suggested, the American coal exporters should prove unable to meet the growing import demands for

Apparent total consumption of primary energy 1950–65 in Europe and the United States

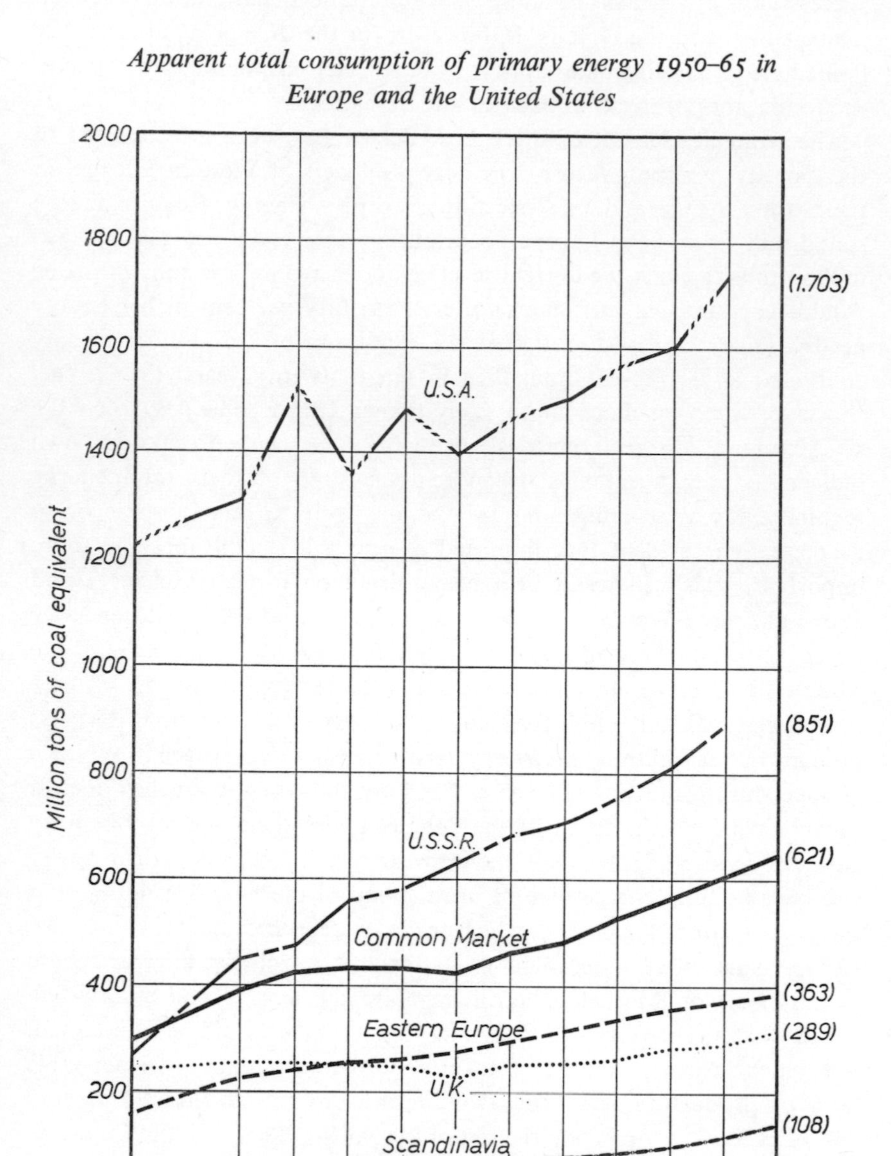

American coking coal all over the world. Such a policy if legalised within the framework of a Community common energy policy could, of course, have important implications for Britain in the event of her accession to the Common Market.

In the shorter-term, one may expect the Community countries to take further measures designed to secure certain markets for coal by means of Government-sponsored agreements between the coal producers and power stations, as in Germany and France, and indirect state assistance, that is, in writing off loans, state contributions to miners' pensions and modernisation expenditure, and encouragement of moderation on the part of the oil companies. Similar measures may conceivably, in due course, be developed in the United Kingdom, particularly if large quantities of natural gas were to be found in the North Sea. One possible solution which was put forward was the suggestion that the National Coal Board should co-operate with the Gas Boards and other interested parties in organising a rational distribution policy along the lines of the solution adopted in Holland; this however, appears to have won little support. Elsewhere in Western Europe the same need for a co-ordinated energy policy with aids for indigenous sources does not apply, either because of the absence of such resources or because the indigenous resources are fully competitive with imported sources of energy, as is the case in Norway and Sweden. It is, however, noteworthy that in Eastern Europe and the Soviet Union the general policy that is being followed is one of developing all available energy resources, even if, in more recent years, the emphasis has been placed on concentrating the greater part of the increase in output on the cheapest and most economic sources. It is therefore in coal production that the energy spectrum of Western and Eastern Europe will differ most sharply in the years ahead, for while in Western Europe coal production will, at best, remain at about its present level, in Eastern Europe it will continue to increase steadily. This, in turn, will accentuate still further the fact that while Western Europe, until the day is reached when nuclear energy can be introduced on a large and economic scale, will become steadily more dependent upon fuel imports as total energy requirements rise, Eastern Europe will continue to be self-sufficient, and may, in time, become a net exporter of energy of considerable significance.

But we must now turn back to Western Europe, and particularly the Common Market countries, where the need for a co-ordinated energy policy is probably more generally accepted than in any other part of Europe, and where the urgency of this problem has been fully recognised. Whatever the final outcome of the great energy debate in the Community, it has been accepted that there is a certain amount of energy, quite distinct from coal, which must necessarily continue to be produced. This is the case either because the price at which it can be produced is exceptionally favourable, i.e. natural gas in the Netherlands, or because it is particularly suitable for a specific form of consumption, i.e. lignite, or because a great deal of heavy expenditure has already been incurred and costs of future production are therefore relatively low. If we add to this a continuing level of coal production in the Community area of some 175 million tons, and accept the total energy consumption estimates of 743 million tons of coal equivalent in 1970 and 1,130 million tons by 1980, then, as we have already seen, import requirements by 1980 may be expected to amount to over 600 million tons per year, or some 60 % of the total needs in energy of the Common Market countries. This is a dangerous degree of dependence.

In conclusion, and fully accepting the fact that forecasts are hazardous and are seldom proved correct, the likely development of energy policy in Europe up to 1980, would appear to be as follows: in the Common Market countries the gradual adoption and application of an undisguisedly cheap energy policy as a result of a freely recognised and accepted dependence on cheap imports from outside Europe. In the United Kingdom coal will no doubt remain the country's main source of energy, but oil, natural gas and atomic energy will certainly steadily, and perhaps even spectacularly, increase their shares of the market. Elsewhere in Western Europe, an undisguisedly cheap energy policy will certainly be followed, based on competitive indigenous resources, that is, hydro-power, and cheap imports of oil and, where necessary or financially worth-while, American coal. In Eastern Europe, a steady development of all sources of energy seems likely, with coal, however, falling steadily behind oil.

Nuclear energy will be developed in all countries, but it is unlikely to emerge from the development stage until after 1970, and perhaps not until after 1975. Its future as a major source of energy in a fuel-starved continent is assured, but its development as a fully economic and competitive contender for the energy markets may not occur until after the end of the next decade. After 1980 atomic energy can be expected to expand at a rate that will make it the dominant source of energy in most parts of Europe. Already, Community experts have estimated atomic energy share of the Community's energy market in the year 2,000 at 30 %; there is no reason to suppose that it will not become at least as important in the United Kingdom.

While Europe, therefore, is today moving towards a multi-fuel economy, this process will become increasingly obvious and a full economic reality only towards the end of the period under review. While the basis for this development has been or still is being laid today, the energy scene in the Common Market countries during the next fifteen years will continue to be dominated by oil, and in the United Kingdom and Eastern Europe, by coal and oil.

INDEX

INDEX